FUN WITH SCIENCE

MOVEMENT

BRENDA WALPOLE

Contents

Use the symbols below to help you identify the three kinds of practical activities in this book.

EXPERIMENTS

TRICKS

THINGS TO MAKE

Illustrated by Kuo Kang Chen · Peter Bull

Warwick Press
New York/London/Toronto/Sydney
1987

Introduction

Pushing, pulling, lifting, stretching, twisting, and spinning are just some of the different types of movement you can explore in this book. Objects cannot move by themselves; they need a force to push or pull them before they can start or stop moving. It takes more force to make things start or stop than it does to keep them moving.

A natural force called gravity pulls things down to the ground but people have invented a number of machines to make things move in different directions and at different speeds. Rollers, wheels, levers, pulleys, and gears all make it easier to move a heavy load. Even the most complex machines have simple levers and wheels somewhere inside them.

The questions on these two pages are based on some of the scientific ideas explained in this book. As you carry out the experiments and tricks you will be able to answer these questions and understand more about how things move in the world around you.

This book covers seven main topics:

- Gravity and weight
- Balancing
- Inertia
- Friction
- Slopes, wheels, pulleys, and levers
- Different types of movement
- Machines and movement

A blue line (like the one around the edge of these two pages) indicates the start of a new topic.

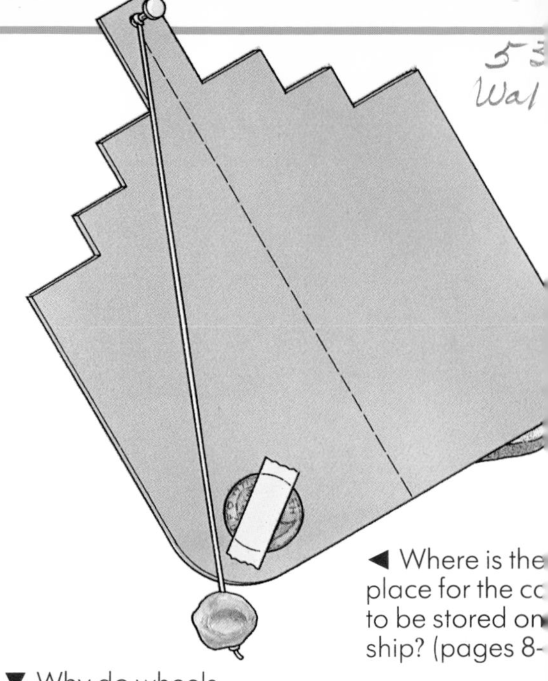

◄ Where is the place for the c to be stored on ship? (pages 8-

▼ Why do wheels make it easier to move quickly? (p. 22–23)

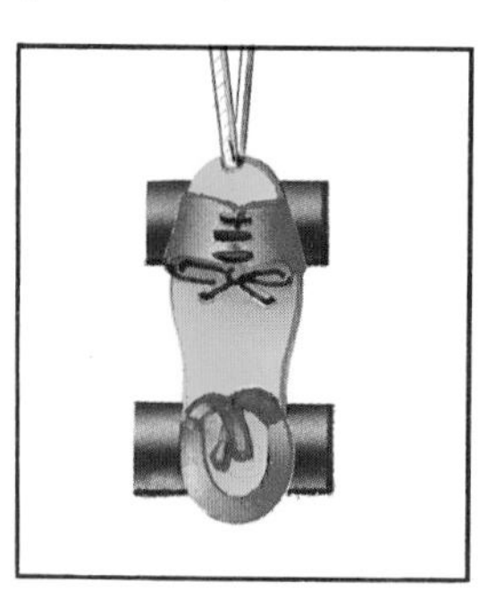

► Why do pulleys make it easier to lift a heavy weight? (pages 24–25)

▼ Why do bowling alleys have a smooth, shiny floor? (pages 16–19)

How does a water
eel produce
ough power to
nd wheat into flour?
ages 36–37)

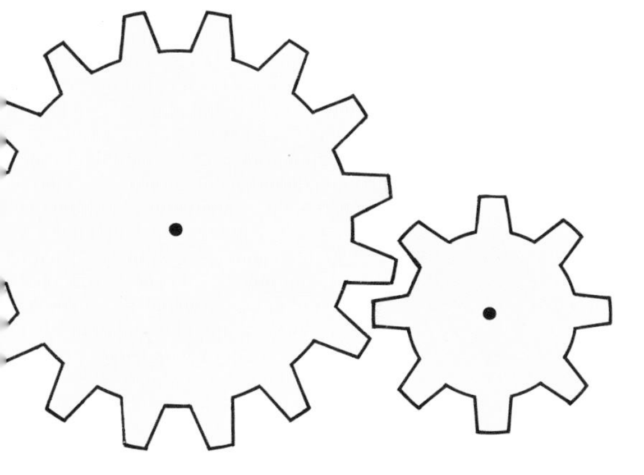

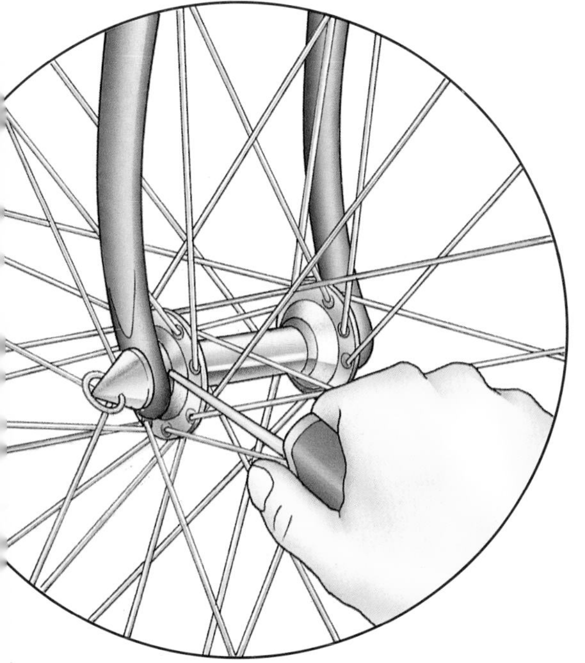

▲ Why do apples fall **down** from the tree? (pages 4–5)

▼ What makes a seesaw balance? (pages 10–11)

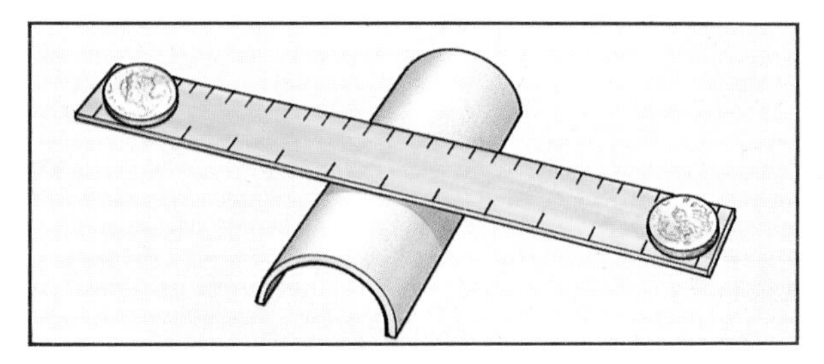

◀ Why do gear wheels make machines move quickly and easily? (page 23)

◀ Why does oiling a bicycle make it go faster? (pages 18–19)

▼ Why do gloves make it easier for a goalkeeper to hold onto the ball? (page 17)

Down to Earth

When an apple falls from a tree, why does it fall **down** to the ground? A famous scientist called Isaac Newton puzzled over this problem while sitting in an orchard many years ago. He suggested that the apple and the earth both had an invisible force that pulled other objects toward them. But the earth was so large and had such a powerful force it was able to pull the apple down to the ground. This force around objects is called **gravity**.

Investigating Falling

In the 1590s, a scientist called Galileo put forward the theory that all objects are pulled down to earth at the same speed no matter what they weigh.

Try to prove Galileo's theory for yourself. All you need is a heavy ball bearing and a marble, a metal tray and a chair. Place the tray on the floor and stand on the chair above the tray. Hold the ball bearing in one hand and the marble in the other. Hold your arms as high as you can and drop the objects down on to the tray. (Try to let go of both objects at the same time.) Listen for the sound of them hitting the tray. Which one lands first?

How it works

You should find that they land together. Gravity pulls them down to earth at the same speed, even though one is heavier than the other.

More things to try

Use different pairs of objects to test the theory. For example: a light, sponge ball and a tennis ball or a cube of sugar and a dice. Choose objects that are the same size and shape but different weights.

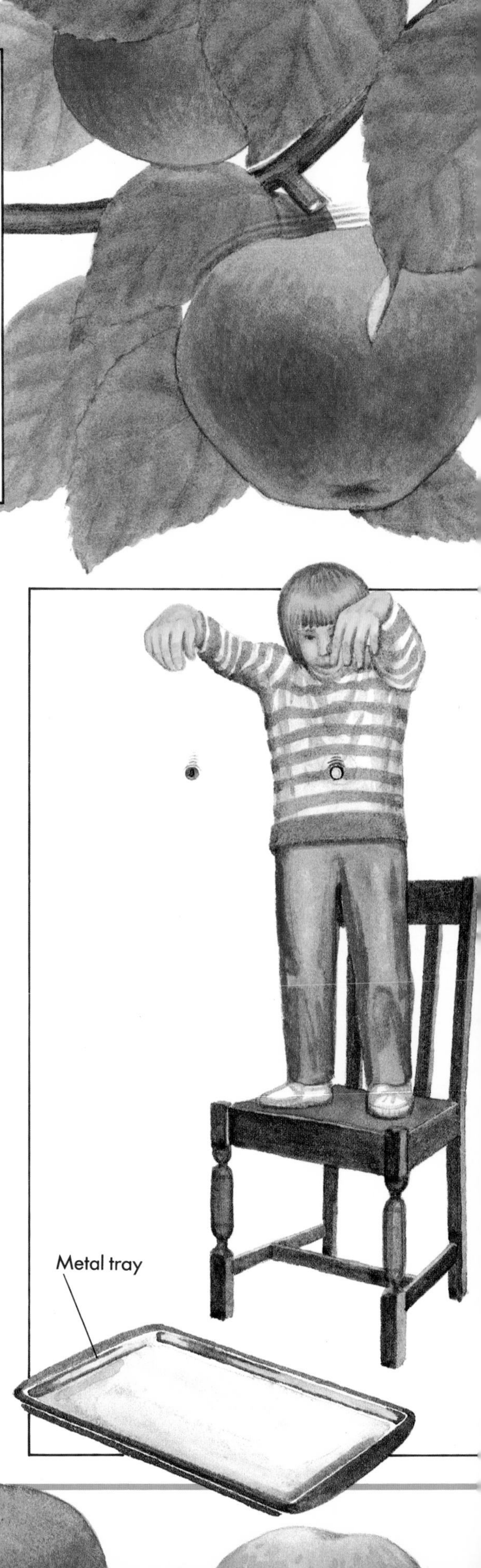

Hitting the Ground

Equipment: Soft modeling clay, a marble or ball bearing, ruler, metal tray.

'hen objects fall, they hit the ground with a ud! The further objects fall, the bigger the ud. Test this for yourself.

Make your modeling clay into a thick, flat ancake shape and put it on to the metal tray.

Drop the ball bearing or marble into the odeling clay from different heights. Try 1 foot 0 cm), 2 feet (60 cm) and so on.

Measure the size of the dent in the modeling ay each time.

Make a chart of your results. What happens s you drop the object from greater heights?

ow it works

bjects that fall from a greater height are aveling faster when they hit the ground than ojects that fall only a short distance. So the oject dropped from the highest point makes e largest dent in the modeling clay. Objects ent the ground if it is softer than they are. But oft objects can be damaged if they hit a hard urface. Think about a soft peach falling from a opping bag!

Falling Coins

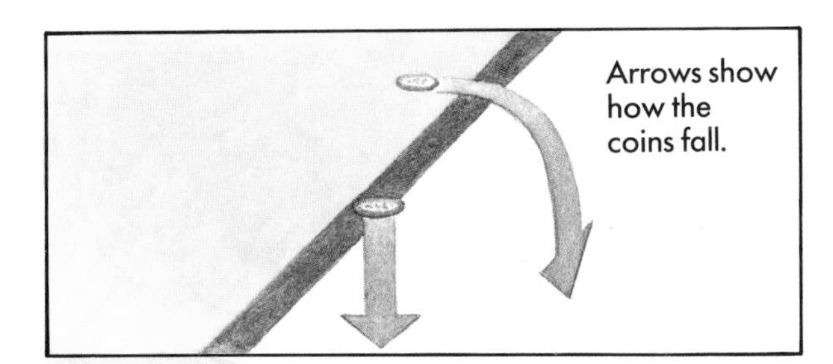
Arrows show how the coins fall.

ce a ruler on the edge of a table so one end t sticks out over the edge and the other end is out 1 inch (3 cm) from the edge. Place two ntical coins in the positions shown in the gram. Use another ruler to hit the ruler on the le and watch carefully to see which coin hits ground first. (You may have to do this several es.)

w it works

th coins hit the ground at the same time despite fact that the coin on the end of the ruler falls aight down while the other coin travels a ger, curved path. This is because the coin that vels further speeds up over the longer distance d catches up with the other coin by the time y hit the ground.

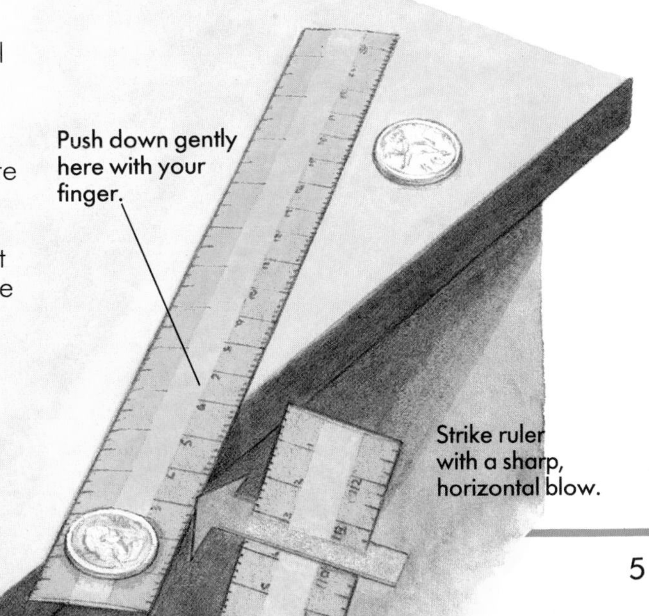
Push down gently here with your finger.
Strike ruler with a sharp, horizontal blow.

Gravity and Weight

Objects have weight because gravity pulls on them. The greater the pull of gravity on an object, the more it weighs.

Make a Spring Balance

This spring balance will help you to compare the weight of small objects.

Equipment: A yogurt pot, thin string, paperclips, a small nail or thumbtack, paper, ruler, pencils, rubber band.

1. Hammer the nail or press the thumbtack into a vertical surface from which you can hang the balance.

2. Loop the rubber band inside the paper clip and hang the paper clip from the nail or thumbtack.
3. Make three holes in the rim of the yogurt pot and thread the string through them to make a handle. Tie the ends together and then tie them on to the end of the rubber band.
4. Make a scale for your balance, using a piece of paper or card fixed behind the rubber band. Mark the point at the end of the rubber band before you weigh something and then mark how far down this point comes when you have something in the pot.

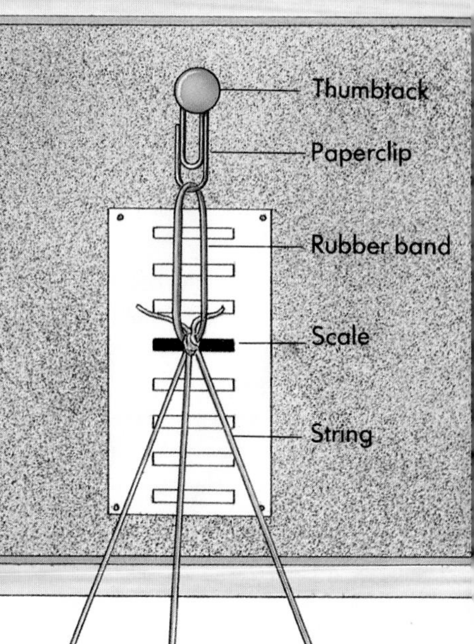

Compare the weight of small objects such as a pencil, a marble, a stone, a handkerchief, or a grape.

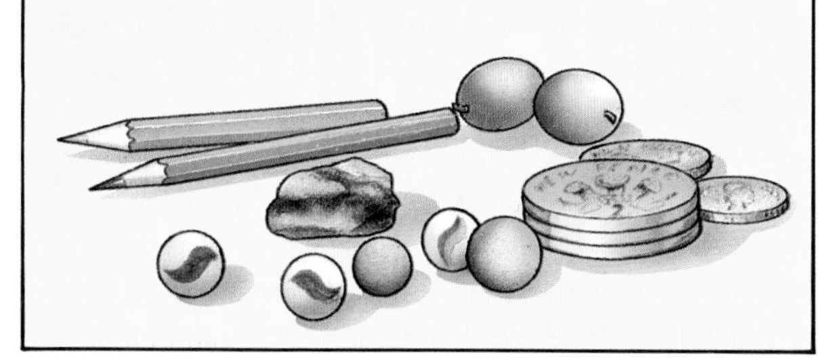

Floating in Space

People do not feel their weight if there is no gravity pulling on them or if they are floating freely. When you bounce on a trampoline you f weightless when you are up in the air but the feeling will last only until you come down to ear again.

The pull of the earth's gravity gets less farther out in space, so things weigh less in space. Astronauts float about in their spacecraft becau there is little gravity to pull them down.

► Astronaut Irwin on the moon during the Apo 15 Mission. Gravity on the moon is about one-sixth of the gravity on earth, so a spacesuit that weighs 183 pounds (83 kilograms) on earth weighs only 31 lbs (14 kilograms) on the moon. This makes it much easier for astronauts to mov about on the moon; they can even hop around like kangaroos!

oving the Oceans

es in the oceans on earth are caused by the
l of the gravity of the moon and the sun.
cause the moon is closer to the earth than the
sun, it pulls the oceans more than the sun. In most parts of the world, the sea level rises about twice a day (high tides) and falls in between (low tides).

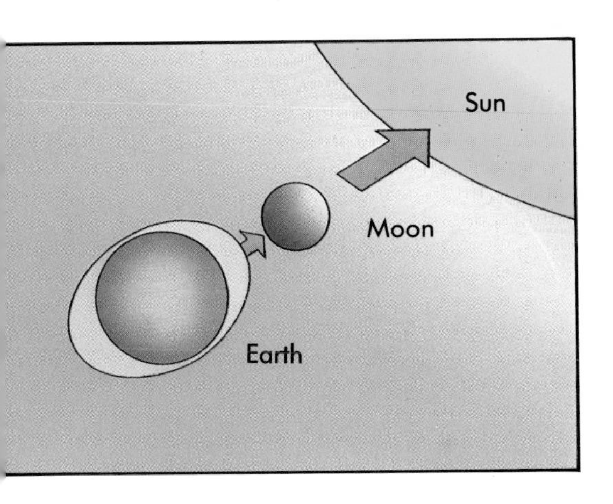

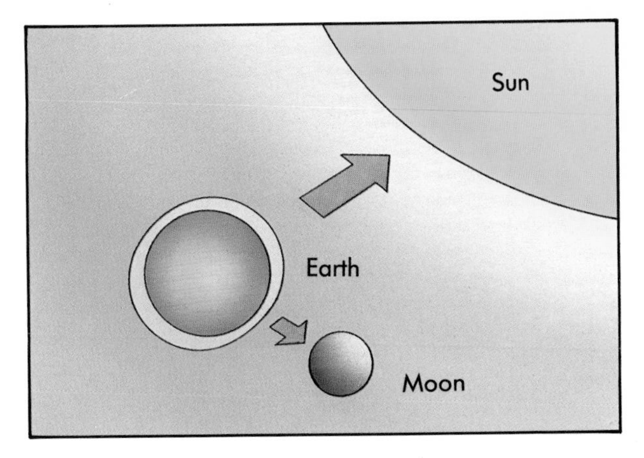

hen the sun is in line with the moon and the
rth their forces of gravity pull together. This pull
akes the tides very high in the part of the earth
osest to the moon. These high tides are called
ring tides. They usually occur about every two
eeks, at the time of a full moon and a new moon.

When the sun and moon are at right angles to each other, their pull is weaker and causes much smaller tides on the earth. These are called **neap tides**. If you live near the sea, you should be able to find out when the spring and neap tides occur in your area.

Balancing

Rest a book on the edge of a table and gradually ease it over the edge. It will balance with part of the book off the table until you push it too far and upset the balance. All objects have a point where they are held in balance by the force of gravity. This balancing point is called the **center of gravity** because it is the place where the whole weight of the objects seems to center.

Find the Balancing Point

The balancing point of a regular shape such as a square or a circle is in the center. This experiment will show you how to find the balancing points of more irregular shapes. All you need is cardboard, string, a weight, and a pin or nail.

Cut out an irregular shape from the cardboard and make three holes in the edge. Tie the weight to the string. When you hold up the string, the weight will make it hang straight down in a vertical line. This is a **plumb line**.

Hang your shape and the plumb line on a pin or nail and draw a straight line down the string. Do the same thing with the other two holes. The balancing point is where the three lines cross.

More things to try

Draw and cut out a boat shape and find the balancing point. Then tape a weight in different places on the boat. How does the weight change the balancing point? Where do you think is the best place for the cargo to be stored on a real ship?

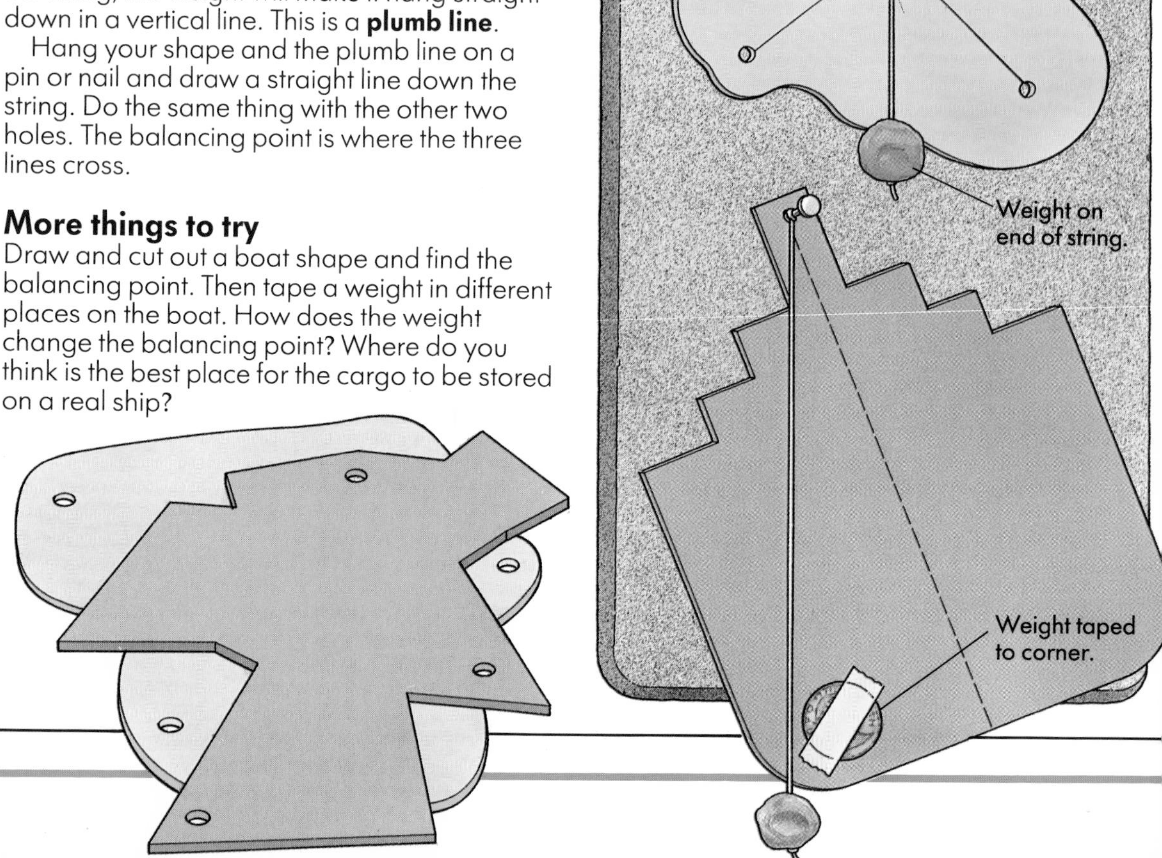
Lines cross at center.
Weight on end of string.
Weight taped to corner.

The Magic Box

ox is a regular shape so you would expect the
ancing point to be in the middle. This trick will
w you how to defy the laws of balancing and
prise your friends. You need a small box and a
vy weight.
ape the weight into one corner of the box.
n put on the lid and show it to your friends.
en the lid away from the weight and let them
the box looks empty. (You could make a false
tom for the box to hide the weight.) Tell them
the box is magic and you can balance it on
Then place the box on a table and gradually
e it off the edge. If you leave the corner with
weight in it on the table, the rest of the box will
g in the air as if by magic!

Heavy weight keeps balancing point in corner of box.

Make a Candle Seesaw

Scrape away some wax from the flat end of
candle so the wick pokes through and you
n light it at both ends.
Measure the candle to find the middle and
sh in the two nails, one on each side.
Rest the nails on the cans to make a seesaw.
Place the seesaw on the tray and check that
balances. Then light both ends.

Equipment: A metal tray, two cans, two thin nails or pins, a long candle.

How it works
Before the candle is lit, the balancing point is in the middle. When a drop of wax falls from one end, the balancing point moves to the other side and the seesaw tips. If the candle drips first from one end and then the other, the seesaw will go up and down as the balancing point moves from one side to the other.

Varning:
emember to put
ut the candle
hen you have
nished.

Objects can balance when their center of gravity allows them to stay upright or poised in position. Here are two balancing toys you can make. They both have their center of gravity low down so it is hard to make them lose their balance.

Make a Tightrope Walker

Equipment: A small potato, a wooden cocktail stick, two forks, thin wire or strong thread.

1. Fix a small potato on the end of a cocktail stick and attach two forks to the potato as shown in the illustration.

2. Make a notch in the end of the cocktail stick so it will fit on the tightrope.

Notch in end of cocktail stick.

Make a Gymnast

1. Carefully draw the shape of the gymnast onto the card or paper.

2. Cut out two shapes exactly the same. One will be the front of your gymnast and the other will be the back. Give your figure a colorful costume.

3. Fix one coin behind each hand using sticky tape. Then stick the two halves of the figure together.

4. When your gymnast is dry, it will balance on its nose almost anywhere. Try balancing it on your finger, the rim of a glass, or a piece of string stretched tightly.

How it works

Although the figure looks heavier at the top, the weight of the coins keeps the center of gravity under the nose so it will balance.

Equipment: Thin cardboard or thick paper, scissors, two small coins (the same weight), glue, sticky tape.

Tape coins to
back of hands.

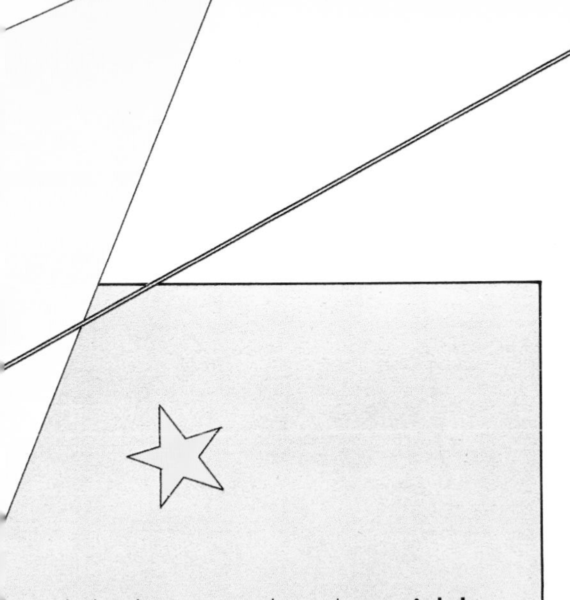

Stretch the thin wire or thread **very tightly** to ke the tightrope.
Balance your cocktail stick on the tightrope. two forks keep the center of gravity below tightrope, which helps it to balance.
Once you have steadied your "tightrope ker," try blowing gently to make it move ng the wire. If the movements are not very ooth, grease the wire or thread or stretch it n angle.

Seesaws

Two people who weigh about the same can balance if they sit on either end of a see-saw. But what happens if one person is much heavier than the other? Try this investigation to find out how to make the seesaw balance.

1. Cut the tube in half, place it flat side down and balance the ruler across it.
2. Put one coin at each end – they balance because the center of gravity is in the middle.
3. Add a second coin to one end. This makes the heavier end lower but can you make the seesaw balance again without adding any more weights?
4. Move the pile of two coins closer to the middle of the ruler until the seesaw balances again. You should find that the seesaw will balance when the two coins are halfway between the center and the end of the ruler. This is because they are twice as heavy as the load at the lighter end. Can you think of another way to make the seesaw balance?

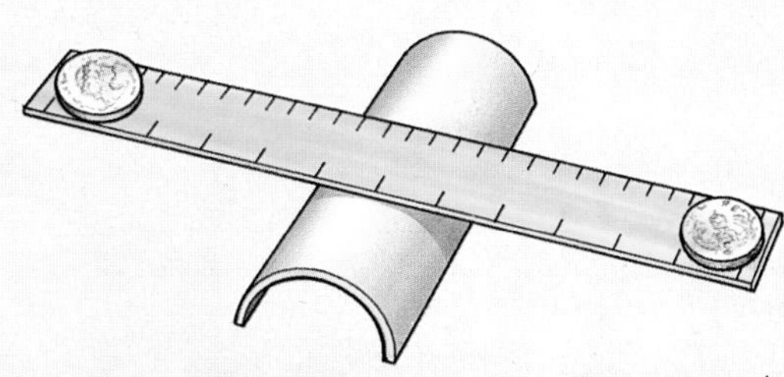

Equipment: A 1 foot (30 cm) ruler, several coins of the same size, a cardboard tube.

▲ A heavy person has to sit nearer to the middle of a seesaw to balance a lighter person at the other end.

Start and Stop

Objects that are still do not want to move and objects that are moving do not want to stop. This tendency of something to stay still or keep moving is called **inertia**. (The word comes from the Latin word for "lazy.") To make something start or stop moving you must overcome its inertia. You can do this by pushing or pulling the object. These pushes and pulls are known as **forces**. The heavier something is the more force it needs to start or stop it moving.

Getting Things Moving

Is it easier to start something moving quickly or slowly? Try this experiment to find out.

1. Tie a length of thread around two heavy books.
2. Rest a board across two empty cans and put the books on top.
3. Gently pull the thread. The books should start moving quite easily.
4. Now keep the thread slack and give it a really hard tug. This time the thread should break because the books have too much inertia to start moving quickly.

Does it need more pulling power to start an object moving . . . or to keep it going?

All you need is a toy car and a rubber band. First try pulling the rubber band. Notice that the harder you pull, the longer it becomes. Then fix the rubber band to the front of the car and pull the car along. You will find that you have to pull quite hard to start the car moving but it needs less pull to keep it going.

Pull
gently
Tug
really hard
Longer band mean
you are pulling
harder.

Spinning Egg Puzzle

ow can you tell the difference between a raw
g and a cooked egg without breaking them?
ertia can help you to solve this puzzle.

in each egg in turn on a plate. The egg that
eps spinning for longer is the cooked one.
ow spin the eggs again but quickly stop them
inning. Then immediately let them both go
ain. The cooked egg will stay still but the raw
g will start spinning again.

ow it works

e contents of the egg have more inertia when
ey are liquid (in the raw egg) than when they
e solid (in the cooked egg). This slows the
w egg down so it stops spinning before the
oked egg. But when you stopped the eggs
d then let go, the liquid in the raw egg was
l moving. This movement started the egg
inning again.

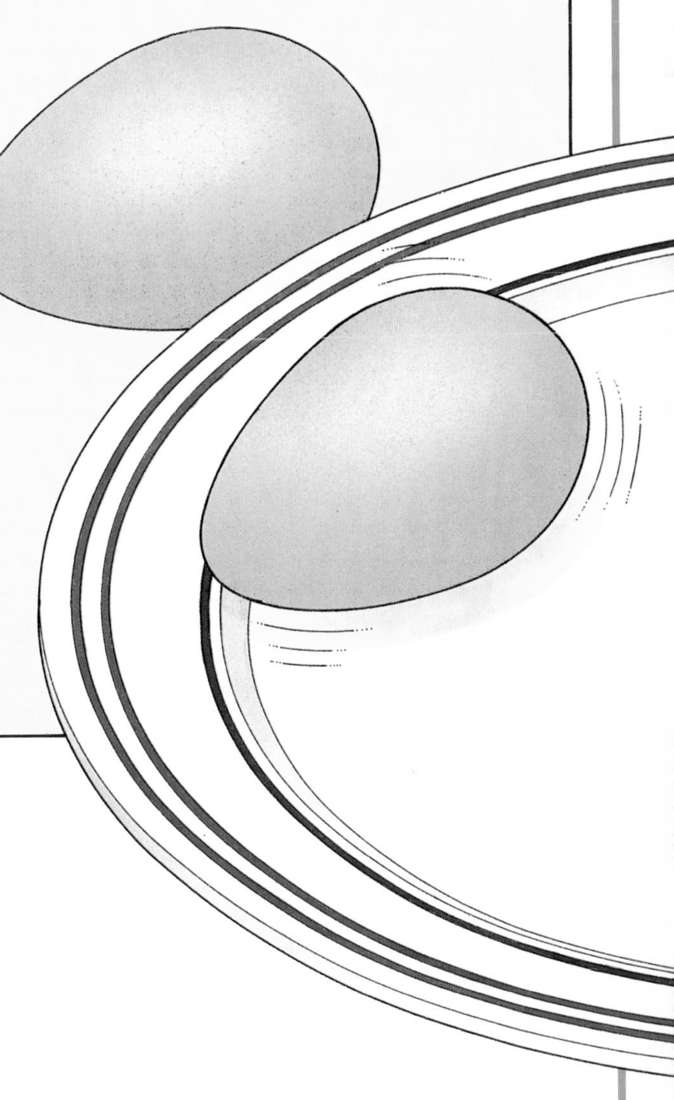

Next time you are in a car, notice what happens if the driver pulls away suddenly. Your inertia pushes you back into your seat – you are not moving and your body wants to stay still. If the driver stops suddenly, you will continue forward as your inertia resists stopping – your body does not want to stop moving.

Seat belts help to overcome your inertia and hold you firmly in your seat. The photographs to the left show dummies being used to test seat belts at a road research center. The dummy in the top picture is wearing a seat belt; the one in the bottom picture is not.

The Lazy Coin

Equipment: A glass, a piece of card, a coin.

Balance the card across the top of the glass and balance the coin in the middle of the card. Can you make the coin fall straight down into the glass without touching the coin?

How it works

Once again you can use inertia to make this trick work. If you flick the card forward, the coin has too much inertia to move and should fall neatly into the glass.

More things to try

- Balance the coin on top of the card on the edge of the glass.
- Use a marble instead of a coin. You will need to put a little sugar or a sponge in the glass to stop the marble breaking the glass.
- Balance a cooked egg on the edge of the outside of a matchbox and put it in the middle of the card. (You will need sugar or a sponge in the glass again.) Only the egg has enough inertia to stay still and so it falls into the glass.

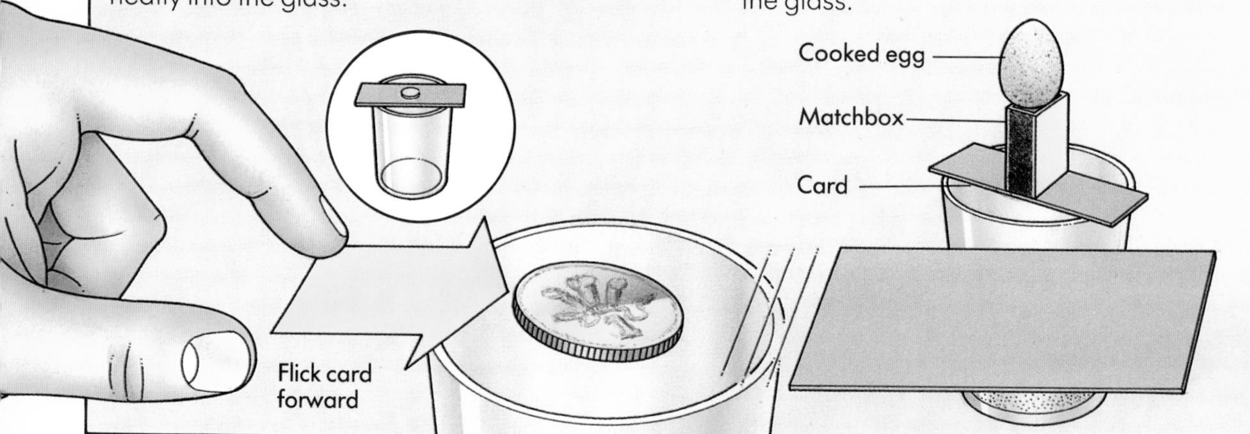

The Tablecloth Trick

You may have seen a magician pull away a tablecloth but leave the cups and saucers safe and sound on the table. This is not a good trick for you to try at home so here is another version that will not do any damage if it goes wrong! All you need is a small sheet of paper and a plastic mug of water (try a glass or an **old** cup and saucer when you have perfected the trick).

Stand the mug of water on the paper on a table. Make sure the outside of the mug is completely dry – the trick will not work if it is wet. Can you pull out the paper without spilling the water in the mug?

How it works

If you pull the paper with a sharp jerk, the mug should stay where it is. (Do not lift the paper; keep it flat on the table.) The mug has too much inertia to be moved by the sudden jerk.

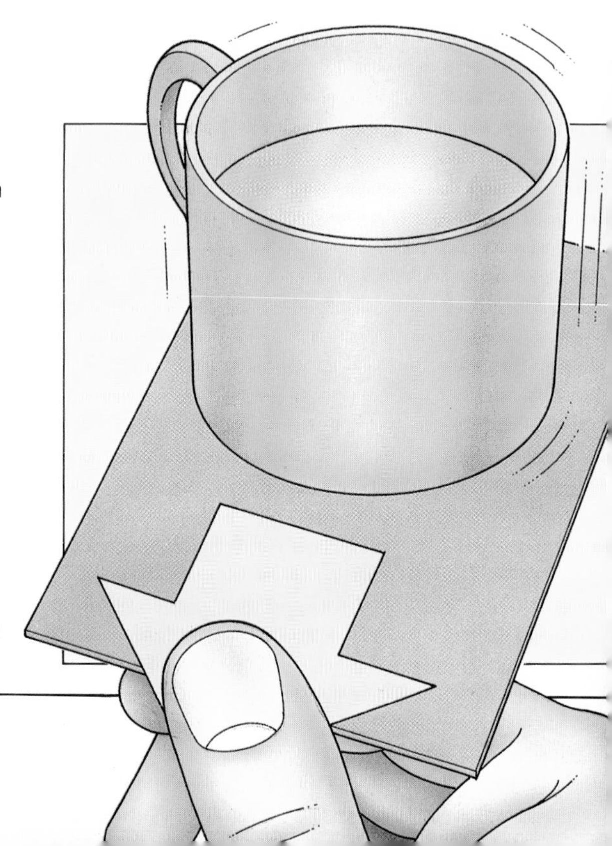

Demolish the Tower

‹e a tower out of checkers on a flat table. Can
take the tower down one layer at a time
›out touching it?

› up the ruler with the bottom checker in the
. Leave the end of the ruler hanging over the
ıe of the table. Then push the ruler toward the
er with a sharp tap or swing it sideways under
tower. As it hits the bottom checker, it should
h it out of the way but leave the rest of the
er standing. With practice, you should be able
emove the checkers one by one.

v it works

tower has a large inertia and the small, sharp
h at the bottom is not enough to overcome this
tia and make the whole tower move.

Sliding Along

One way of moving things is to slide them over another surface. Think about pulling a sled. Does it slide more easily on ice or on a concrete path? When two rough or uneven surfaces rub together an invisible force called **friction** holds them back and makes moving difficult. Moving is easier when there is little friction between two surfaces.

► Smooth or even surfaces produce less friction. That is why it is easy to zoom down a shiny slide in the park.

Investigate Friction

Arrange a selection of objects in a line along the edge of a smooth piece of wood. Then slowly raise the wood until the objects begin to move. Make a note of the objects that move first. Repeat the experiment using the metal tray. Do the objects move more easily . . . or less easily? Do you have to lift the metal tray higher than the wooden board before the objects will move? Which surface has the lowest friction?

How it works
Some of the objects move more easily than others because there is less friction between their outer surface and the surface of the board or tray. Feel the objects that move easily. They should feel smooth.

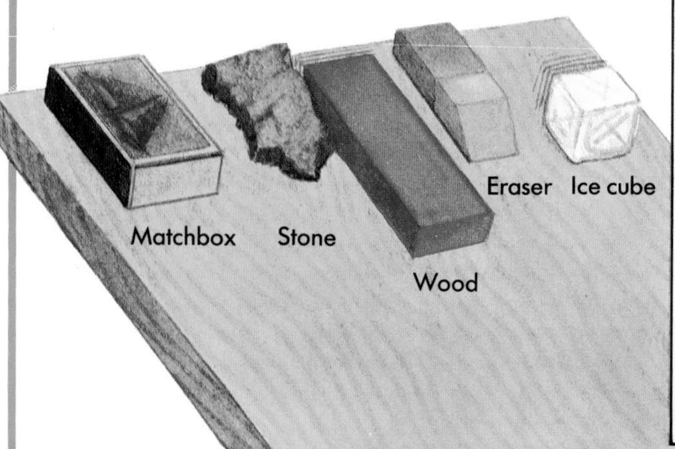

Friction in Water

Friction does not only hold back objects on solid surfaces. It also makes it more difficult fo objects to move in water.

Take a smooth rubber ball and a tennis ball. Put a little water in a shallow bowl. Try spinni each ball in the dish.
Which one moves more easily?

How it works
Once again, the smooth surface causes less friction, so the rubber ball moves more easily than the tennis ball. This is why a fast boat hc a smooth hull.

ction Keeps Things Moving

ion always makes it harder to move things but can sometimes be very useful. For example, riction between the soles of your shoes and ground stops you slipping over when you k, and the wheels of a car could not grip the d without friction. In the illustration below are e examples of how useful friction can be.

studs on soccer boots ease friction to stop the ers slipping over.

ion allows the ers to kick the Without friction it d slide off their

Friction helps to keep the screws in the wood of the goalposts and stops the knots in the netting from coming undone.

Gloves create friction, which helps the goalkeeper to hold onto the ball.

Life Without Friction

Here are some tricks that will show you how difficult life might be without friction.

- Screw on the top of a glass jar as hard as you can. Then wet your hands with soap and water and try to remove the lid. You will find it is impossible! The soap and water reduce the friction so much that you cannot grip the lid well enough to unscrew it.

- Rub a little vaseline or margarine on to the handle of a door (remember to clean it off again afterward). Then try to turn the handle. Once again you will find you need friction to open the door.

Increasing Friction

Sometimes it is very useful to increase the amount of friction between things to keep them moving. For example, in icy conditions grit is spread on roads to make the surface rougher, and increase the friction between the tires and the road. This helps the tires to grip the road. Tractors and snow-plows have tires with large, deep treads or grooves to produce a lot of friction and give them a good grip on slippery surfaces.

Reducing Friction . . .

Smoothing out surfaces helps to reduce friction and make things move more easily. Polishing surfaces helps to keep them smooth. The special floors in bowling alleys are polished for this reason. Water and oil can also be used to reduce friction. They fill in some of the bumps in a surface or form a layer that stops two surfaces from rubbing against one another.

. . . With Water

Equipment: A smooth metal tray, books, a small flat bottle, water, soap.

1. Prop up the tray on the books to make a slope.
2. Wet one side of the tray and try sliding the bottle down each side in turn.
3. Now rub soap on the wet side and slide the bottle down again. On which surface does the bottle slide most easily?

How it works
There is most friction between the glass and the dry metal of the tray. Even though they feel smooth, there are bumps in the glass and the metal. The water fills in some of the bumps in the surfaces so there is less friction. The soap fills in even more bumps and the bottle slides very easily. In fact the bottle slides on a layer of soapy water, not on the metal. Wet things are slippery because water smooths out the bumps in surfaces. This can be dangerous – it is easy for a car to skid on a wet road.

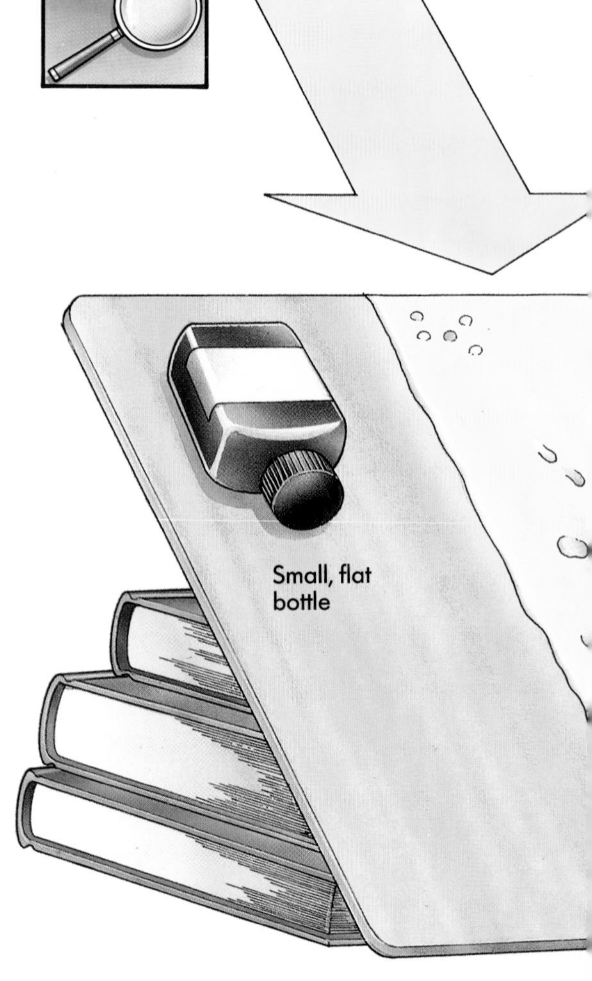

smooth, polished floor of a
vling alley and the shiny surfaces
ne balls help to reduce friction.
; makes the balls roll down the
vling lanes very quickly.

. With Oil and Grease

ɔther way to reduce the friction between two
faces is with oil or grease. Oil or grease do not
up like water so they work for longer.

you need are two pieces of sandpaper and
ne vaseline or cream. Try rubbing the pieces of
ıdpaper against one another. It is hard work
t it? Now spread a layer of vaseline or cream
the surface of one piece of sandpaper and try
ain. You will find that the greasy surfaces slide
ainst each other easily with very little friction.

and grease are used in machines such as
ycles or cars to reduce the amount of friction.
oil or grease forms protective layers
ween the moving parts and stops them
bing against one another.

Oiling a
bicycle wheel

Slopes and Rollers

Heavy things are very difficult to move but there are several ways of making them easier to move. You can find out more on the next eight pages.

► In this exit to a multi-story car park, the vehicles drive down a slope that winds around and around. This allows them to descend easily from a great height within a small space.

Investigate Slopes

Equipment: A heavy toy with wheels (such as a roller skate), string, a plank, a chair.

Make a loop out of the string and tie it to the toy. Put your finger through the loop and try to lift the toy to a height of about 2 feet (60 cm). You will find it is hard work!

Then use the plank to make a slope up to the seat of the chair. Put your finger through the loop of string again and pull the toy up the slope. You will find it is easier to pull something up a gentle slope than a steep one.

How many examples can you find of slopes or ramps being used to make it easier to lift things? Look in garages and railroad stations.

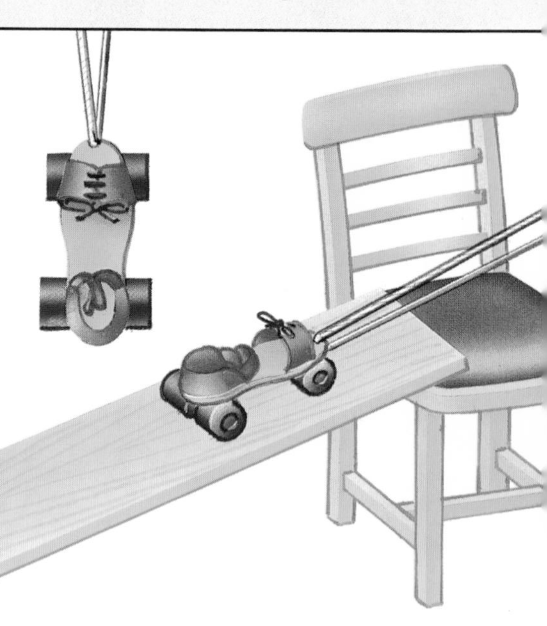

It is easier to move up a gentle slope than to try and climb straight up a steep slope – even though you may travel farther. This is why mountain roads often wind around and around. If the road went straight up the mountain the slope would be too steep for cars and trucks to climb.

► The ancient Egyptians used winding slopes t help them build structures such as pyramids. Th also used rollers made from tree trunks to help them move the heavy stones they needed to bu their monuments.

Rolling Along

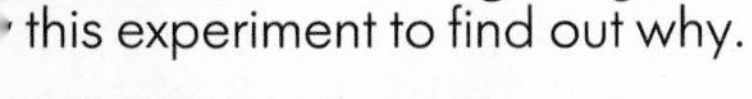

; easier to move a heavy load on rollers her than slide it along the ground or carry it. ' this experiment to find out why.

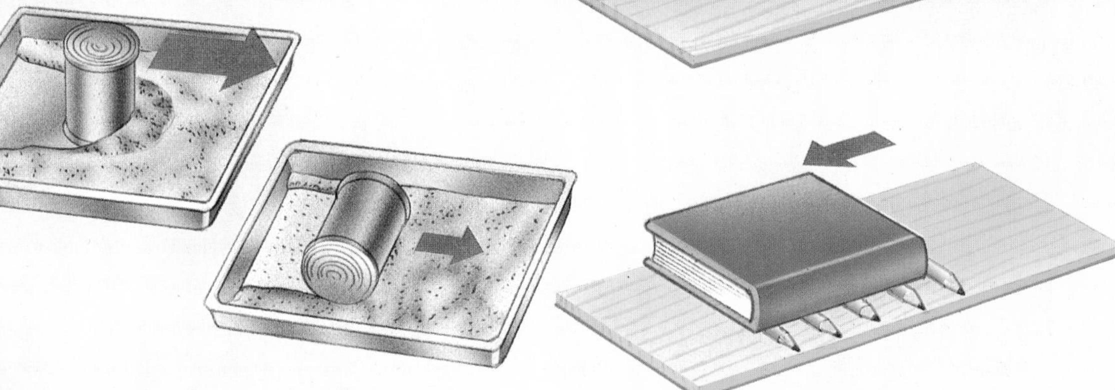

uipment: A metal tray, salt or flour, a can.

Spread out the flour or salt on the tray. Stand the can on its end and push it along. Then turn the can on its side and roll it along.

w it works

you slide the can along, you push the flour salt into little heaps and this makes it more ficult to move the can. Rolling is much easier cause you smooth out the flour so there is s friction.

It is possible to roll objects that are not round by putting rollers underneath them.

Take a heavy book and three or four **round** pencils. Try pushing the book along a table. Then balance the book on the pencils and push it along again. It is much easier to move the book if the pencils act as rollers. But you will have to keep putting the back pencil under the front of the book as it moves forward.

Wheels

Wheels make it easier to move things. They are more useful than rollers because they can be fixed to whatever has to be moved. This makes it possible to move a heavy load quickly and easily.

The wheel was invented about 6,000 years ago but no one knows who invented it. Perhaps it was someone who had used logs as rollers to help them move things. Wheels are made of all sorts of materials such as wood, plastic, metal, or rubber.

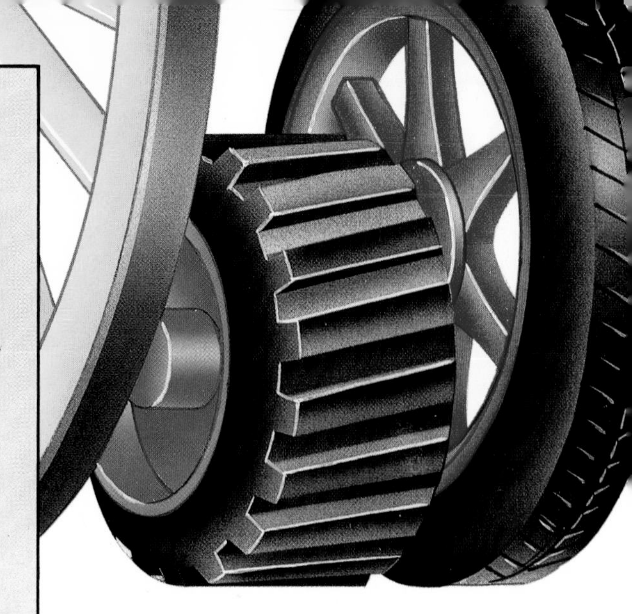

Spin the Book

Many of the wheels on fast-moving machines (such as cars and bicycles) have ball bearings inside them. This trick will show you why. Place a circle of marbles in the rim of the can and balance the book on top. If you push the book gently, you should be able to spin it easily. Try again without the marbles. It can't be done!

How it works
The marbles reduce the friction between the book and the tin so it is possible to spin the book. The ball bearings between a wheel and an axle work in a similar way. In a wheel without ball bearings, the axle and wheel rub together and this slows the wheel down.

Equipment: An empty can with a rim (such as a molasses or cocoa tin), marbles, a book.

king at Gears

:ial toothed wheels of different sizes, which :alled **gears**, can work together to make js work more quickly or more slowly. Each a large wheel turns once it can turn a smaller el several times. The number of times the ller wheel turns depends on the number of or "cogs" each gear wheel has.
ok for gears on machines such as bicycles, <s, or egg beaters. Do the gears speed up or down the parts being turned? Notice how ɔig gear wheel on an egg beater fits into the ller wheel and count how many times the l wheel turns when you turn the big wheel. (If mark the wheels with a marker pen you will ble to count the turns more easily. Wash any <s off the beater afterward.)

t These Gears

e the gear wheels below on to thin board. Push a pin or small nail through the dle of the wheels and fix them to a sheet of board so they will turn round easily. Arrange smallest and largest wheels so that the cogs t.
ɔw many times does the small wheel turn if u turn the big wheel once?
ɔ both wheels turn the same way?
repeat your experiments with three gear els in a row. Try to guess which way the third wheel will turn before you try the experiment.

Changing Direction

Gear wheels can also be used to change movement from one direction to another.

Equipment: Two round slices of potato, several cocktail sticks, two long, thin nails.

1. Fix 6 pieces of cocktail stick into the sides of each potato slice.
2. Push one nail through the middle of each slice to act as an axle.
3. Pin one slice up onto a sheet of cardboard or a notice board.
4. Hold the second wheel up by its axle in a horizontal position and use it to turn the verticle wheel.

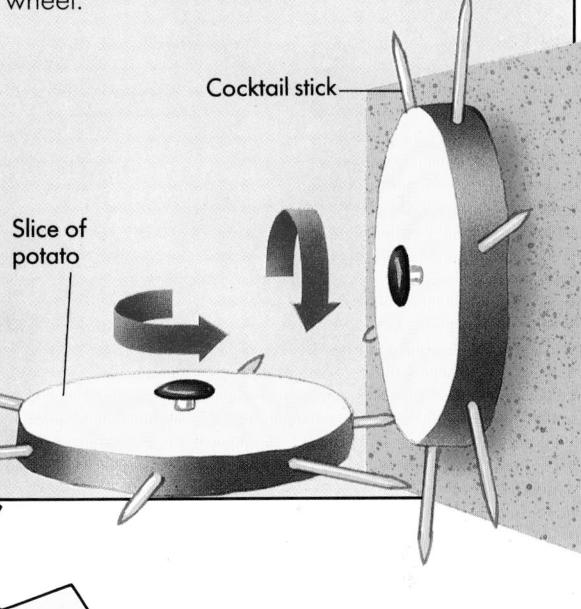

Trace the gear wheels separately, **not** one inside the other.

Arrange the gear wheels like this so the teeth fit together.

Pulley Power

Pulleys are a special sort of wheel. A pulley wheel has a groove all around the rim for a rope to fit into. If you attach one end of the rope to a heavy load you can lift it more easily.

Cranes use pulleys and levers (see pages 26–27) to help them lift heavy loads. Count the number of pulleys on the cranes you see. The biggest cranes have three or four pulleys. A motor provides the power to pull the cable over the pulley wheels.

Brush and Rope Trick

Amaze your friends with your super strength us this simple trick. Ask two or four friends to hold two brush handles apart. Attach a length of ro to one brush and thread it around the two brus as shown in the diagram. Take hold of the free end yourself. Ask your friends to try and keep t brushes apart while you try and pull them together. You should find that you are easily ab to beat the pulling power of your friends.

Hint: Dust the brushes with talcum powder bef you give them to your friends to hold. This will reduce the friction and make it easier for you to pull the brushes together.

Make Your Own Pulleys

1. Bend about 8 inches (20 cm) of wire into a triangle shape and push the ends into a cotto spool. (Ask an adult to help you cut and bend the wire.)
2. Find a suitable place to hang your pulley. hook in the shed or garage or the hook at the end of a plant hanger will do.
3. Tie one end of the string to the handle of t load.
4. Wind the string over the cotton spool.

- Is it easier to lift the load with the pulley?
- How much string do you have to use to lift t load 1 foot (30 cm)?

Now try a double pulley . . .

1. Make two wire triangles. Use about 1 foo inches (35 cm) of wire for each one.
2. Attach two cotton spools to each triangle.
3. Thread the string around the pulleys as shown in the diagram. Use about 6½ feet (2 meters) of string.
4. Attach the heavy load to the pulley as before.

Equipment: Wire, cotton spools, string, a hook, toy bucket full of heavy objects.

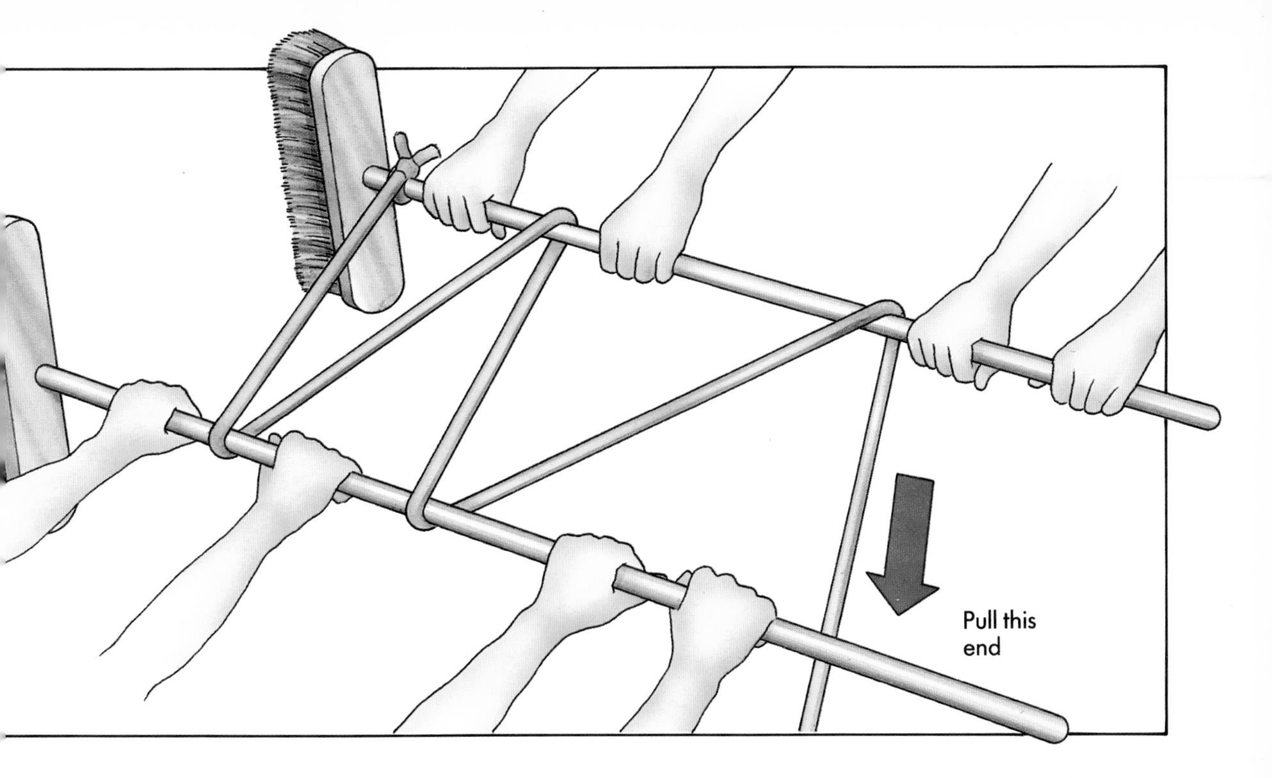

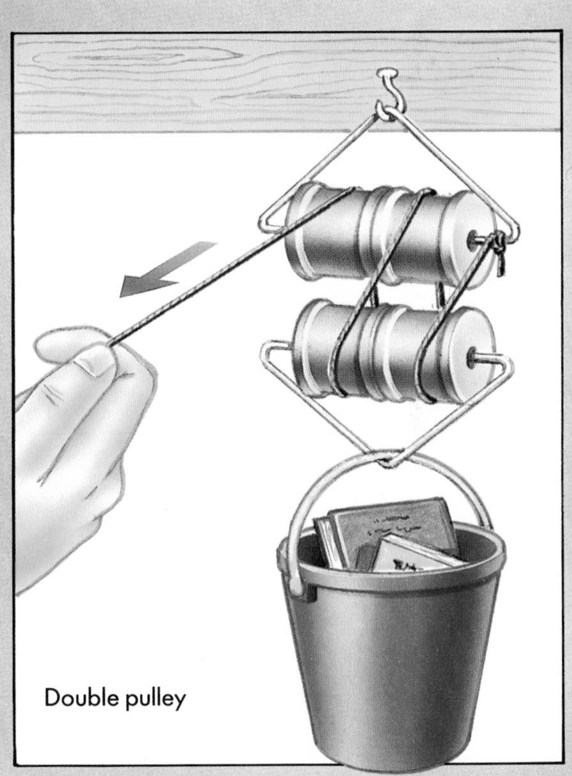

. it easier to lift the load with the double
ulley?
low much string do you need to raise the
ɔad 1 foot (30 cm)?

How it works
The pulley with one cotton spool allows you to lift a heavy load directly underneath the pulley. The double pulley means you use only a quarter of the pull but you need four times the amount of string.

Levers and Lifting

One of the simplest ways of lifting heavy things more easily is to use a lever. Levers work by increasing the pushing force underneath the object so a large load can be moved with a small effort. Levers lift objects most easily when the resting point – the **"fulcrum"** – is close to the object and the pushing point is as far away as possible.

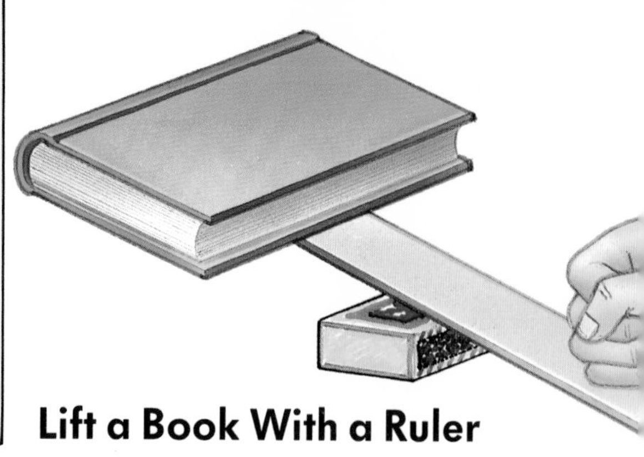

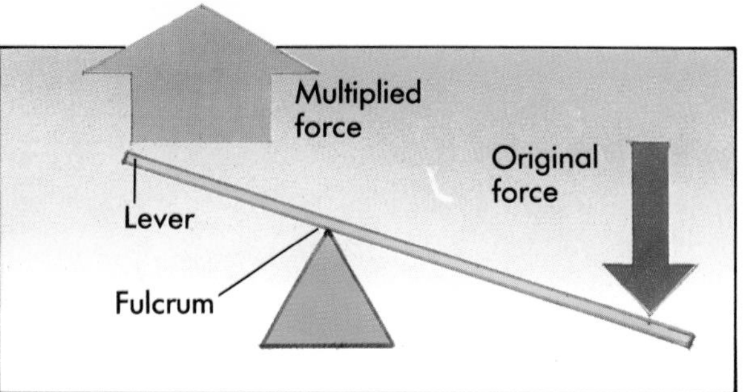

Lift a Book With a Ruler

Choose a heavy book. Lift it up and notice hov heavy it is. Make a lever using a ruler balance across a matchbox. Make sure that the fulcrur (the place where the ruler rests across the mat box) is close to one end of the ruler. Place the book on the end of the ruler nearest to the fulcrum. You will find that you will be able to lift book easily by pressing down gently on the ot end of the ruler. Notice how far down you hav push the ruler and how high the book is lifted.

Jumping Coin Trick

Equipment: A ruler, a pencil, two large coins.

Use this trick to find out where to push on a lever to get the best lift.

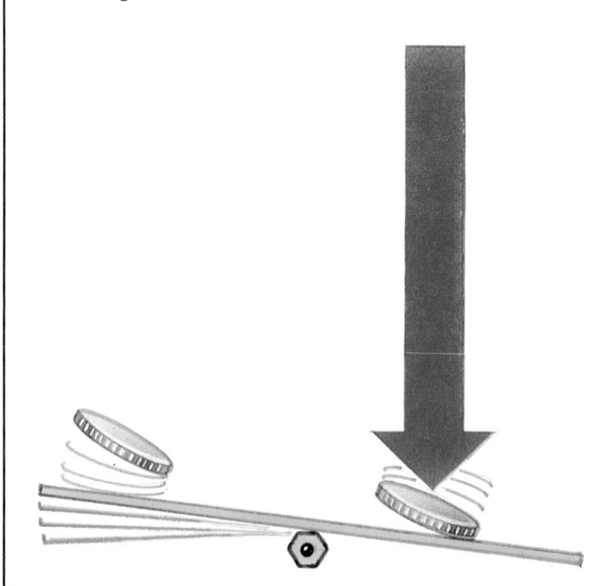

1. Put the pencil under the middle of the ruler and place a coin on one end. Drop the second coin from a height of about 1 foot (30 cm) so it hits the ruler at about the 3 inch (7.5 cm) mark. Notice how high the first coin jumps into the air.

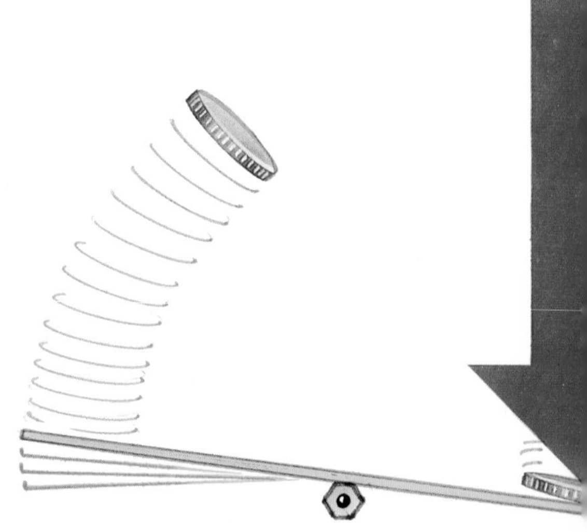

2. Now repeat the trick but drop the second coin right at the end of the ruler. (Be sure to drop it from the same height.) You should see that the first coin jumps much higher into the air this time.

Hit the Outfield

You may be surprised to learn that your arm is a lever too. Throw a ball into the air and hit it with your hand. Then try hitting the ball with a bat. You will find that you can hit the ball with a much greater force using the bat. Your arm is working as a lever with your elbow as the fulcrum. Your muscles provide the pushing force. When you use a bat, the hitting point is further from your elbow (the fulcrum) so the pushing force is greater. The picture shows a cricket bat being used.

obats use the same idea in r tricks. The acrobat who is pring off the seesaw stands e to the fulcrum. The obat who jumps onto the saw to catapult his partner ps as far away from the rum as possible to give the atest possible lift.

w it works

coin hits the ruler with the same force ause you dropped it from the same height. coin jumped higher the second time ause the lever (the ruler) has more lifting ver when the pushing force is further away n the fulcrum.

More things to try

- Try prising the lid from a tin. Is it easier with a short lever or a long lever?
- Where is the best place for a light person to sit on a seesaw in order to lift a heavy person?

Investigating Pendulums

Equipment: String and some weights, a hook, a watch with a second hand.

Cut two lengths of string – make each one about 3 feet (1 meter) long. Tie a small weight to one piece of string and a larger weight to the other piece. Tie each pendulum in turn to a hook or somewhere where it can swing freely. Set the pendulum swinging gently and time how long it takes to swing to and fro ten times. You will find that both pendulums take the same amount of time to complete ten swings even though they have different weights on the end.

Then try some experiments with one weight. First attach it to a long string. How long do ten swings take? Then try again with a shorter string. You will find that the pendulum with the shorter string swings much faster than the one with the longer string.

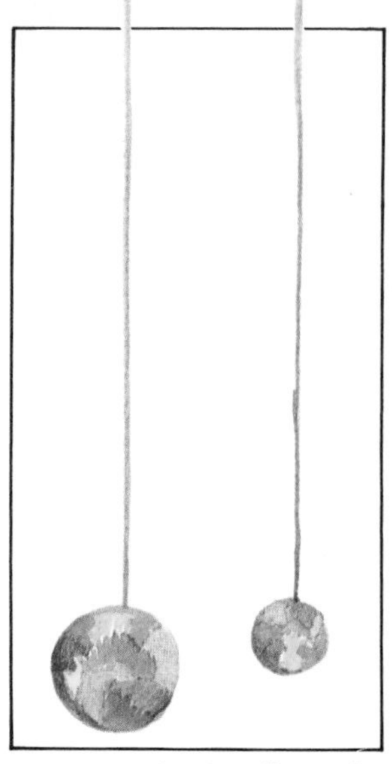

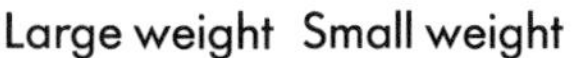

Large weight Small weight

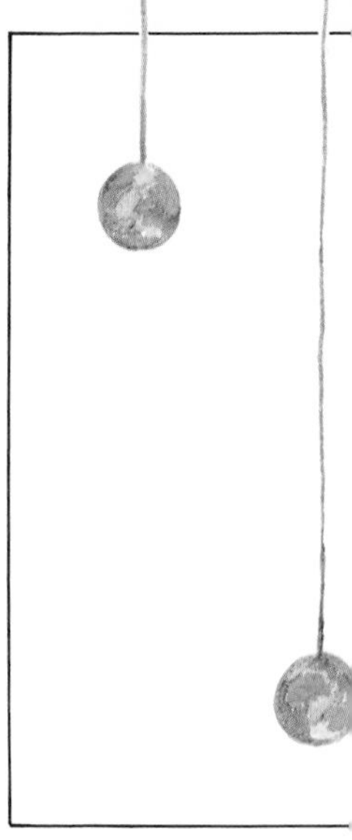

Short string Long st

► Pendulums are used in clocks. They swing to and fro at a fixed rate so they make the clockwork mechanism move at a steady speed.

Swinging

On the next eight pages you can find out about different sorts of movement, such as swinging, twisting, stretching, and spinning. On these two pages you can discover more about swinging by looking at how pendulums work. A pendulum is a rod or string with a weight called a "bob" on the end. In the 16th century Galileo noticed that the chandelier in the cathedral at Pisa took the same time to complete one swing whether the swing was a long one or a short one. He also found that the time of the swing depended on the length of the pendulum – the weight on the end made no difference. Try this yourself.

Shifting Pendulums

e pendulum can set another swinging. Here is
v it works.
ipment: Modeling clay, string, two chairs,
vy books (optional).

two pieces of string about 1½ feet (45 cm)
g and attach a piece of modeling clay to each
ce of string. Tie some string tightly between the
ks of two chairs and put some books on the
irs to hold them steady (or ask someone to
d the chairs). Tie the pendulums to the line of
ng. Hold one pendulum still and start the other
nging. What happens when you let go of the
ond pendulum?

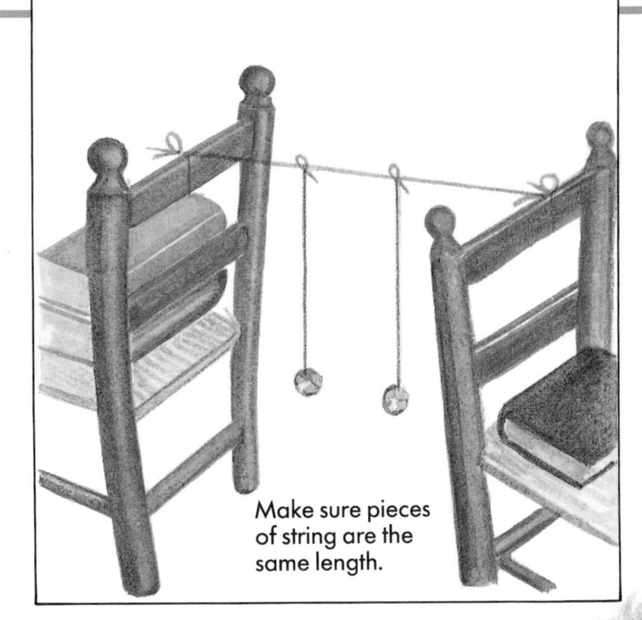

Make sure pieces of string are the same length.

Pendulum Skittles

nis is an outdoor game of skill that you can
ake very easily.

uipment: A ball about the size of a tennis
all, string, a flat board or flat surface, pencils
nd cotton spools (or empty, plastic lemonade
ottles).

Fix about 5 feet (1.5 meters) of string to the
all.
Hang the ball from a suitable tree branch or
her overhanging bar so it swings about 6
ches (15 cm) above the ground.
Make skittles by standing pencils in the
iddle of the cotton spools.
Set up the skittles on the board or on the
round if it is flat.

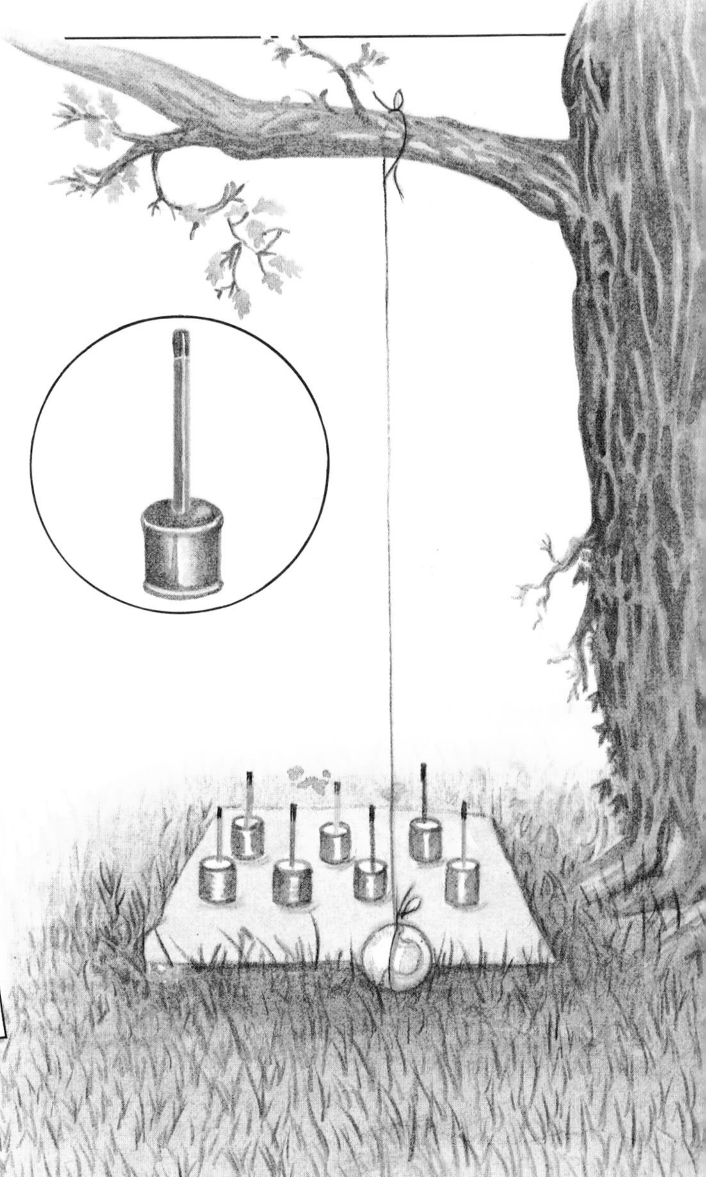

ules
wing the pendulum so it does **not** hit the
skittles on the outward swing but knocks
them over on the return swing. This will take
some practice! You can make the game
more interesting by giving each skittle
different points and keeping the score.

Stretch and Twist

What makes a ball bounce or a catapult spring? Some substances, such as elastic or rubber, stretch when you pull them but spring back to their original shape and size when you let them go. You can use this springy energy to have fun!

Jumping Monsters

Equipment: Thin card, colored pencils, a long rubber band, scissors.

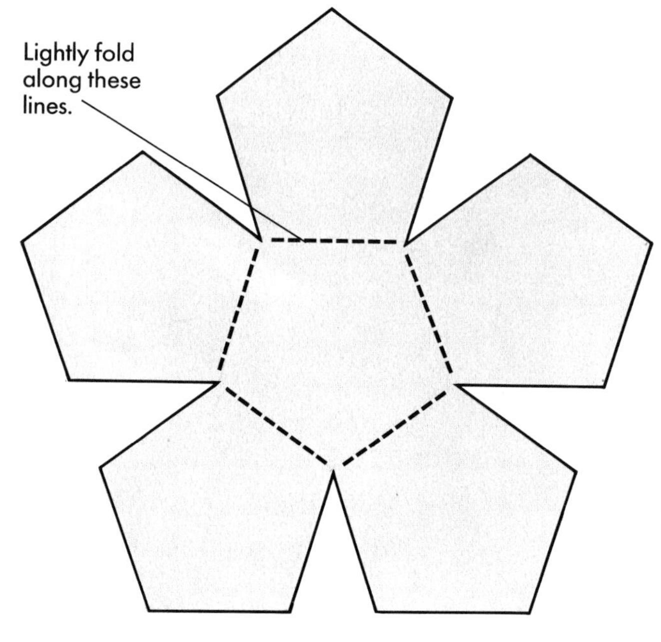
Lightly fold along these lines.

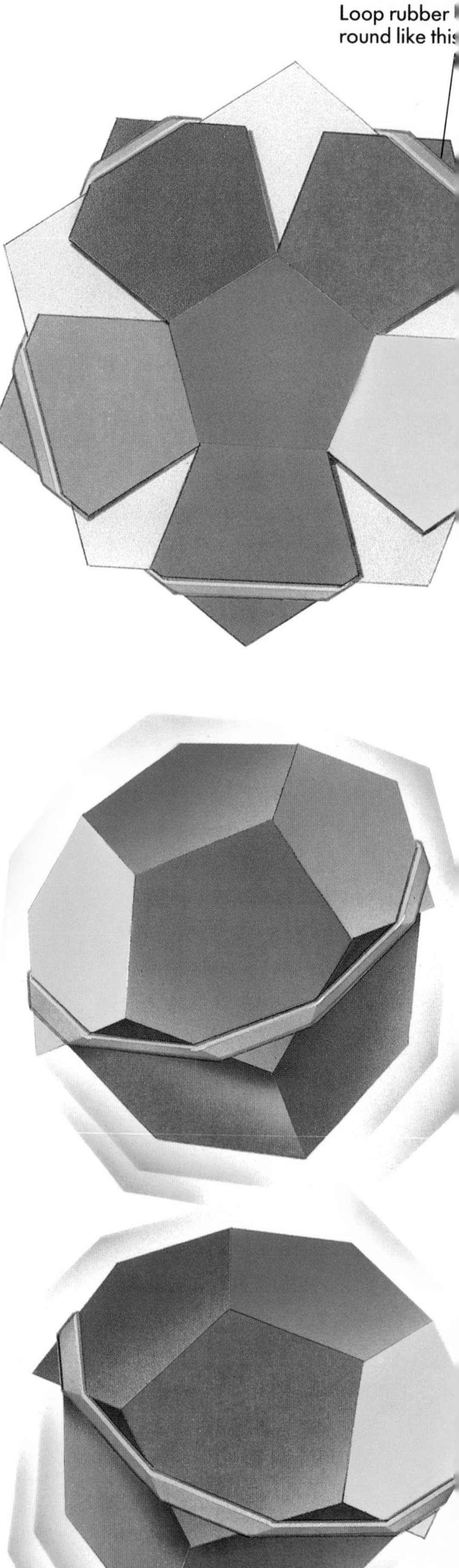
Loop rubber
round like thi

1. Draw a five-sided shape like this onto the thin cardboard. Lightly fold back where shown by the dotted lines.
2. Cut out the shape and then make another one exactly the same.
3. Decorate your monster with patterns or faces.
4. Overlap the two shapes and loop the rubber band over every other corner to hold the two halves of your monster together. Make sure the rubber is slightly stretched but not too tight.
5. When you let go of the monster it will jump up into a solid shape.

How it works
The energy in the stretched rubber band pulls the cardboard into the monster shape.

Magic Rolling Can

uipment: A large can with a lid, a ımmer and nail to make holes, a rubber ınd, a heavy nut or other weight, string.

ːk an adult to help you make two holes in e lid and two holes in the end of the tin ing the hammer and nail.

Cut the rubber band so you have one ng piece.
Thread the band through the holes in the ın so it crosses over in the middle. Knot the ıds of the rubber band together at the lid ıd.
Tie on the weight inside the can.
Press on the lid and roll the can forward. 'hat happens?

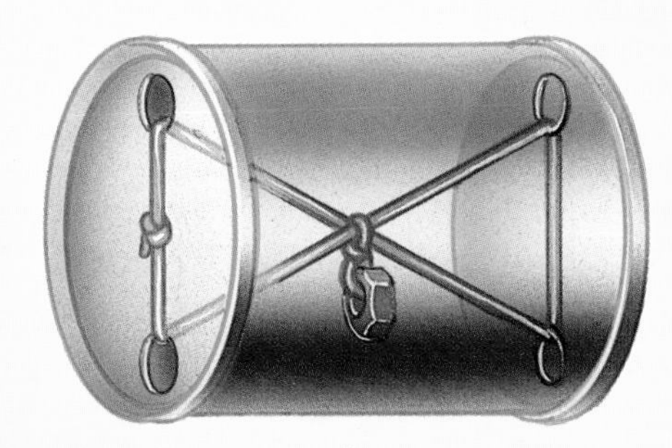

ow it works
ɔu will find you have made an obedient ın, which always comes back to you! This because the heavy weight stays hanging elow the rubber band so the rubber ecomes twisted. (Don't push the can too ard or the weight will spin too.) The can ɔlls back on its own because it is driven long by the energy stored up in the twisted ıbber.

Shrinking Rubber

Weight-lifting is a good trick to try using a rubber band. Most materials expand (get bigger) when they are heated but rubber does the opposite.
1. Cut a rubber band to make one long piece and tie one end onto a toy car or similar weight.
2. Hang the rubber band on a hook so the weight is just resting on a table or other surface.
3. Heat the rubber band by moving a hair drier or a lighted candle up and down a few times. You will see that the weight is lifted a little way off the table.

Warning: Take care if you use a candle. Move the flame fairly quickly so the band does not melt.

How it works
The heat makes the rubber band contract (get shorter) for a little while so it pulls the weight off the table. But look at the elastic band afterward. Does it go back to its original size?

Make a Creeping Crawler

Equipment: A cotton spool, a small rubber band, matches, a candle, sticky tape, penci scissors or penknife.

1. Cut a slice about ½ inch (10 mm) thick from the candle. Make a hole through the middle (where the wick was) using a sharp pencil.
2. Make a groove in one side of the slice using a pencil point or penknife.
3. Push the rubber band through the hole in the slice and place a match through the loop. Pull the rubber band tight so the match fits into the groove.
4. Thread the other end of the rubber band through the hole in the cotton spool.
5. Push half a match through the loop of the rubber band that comes through the spool. Tape the loop and half match firmly to the end of the cotton spool so they cannot turn around.
6. Now wind up your toy by turning the long match (at the candle end) round and round. When you put it down, the toy will start to crawl. You can scare your friends by slipping the toy under a tissue or napkin and making it move as if by magic!

How it works
As you turn the match, you twist and tighten the rubber band. As the band unwinds, it releases the energy stored in the twisted rubber and makes the toy move.

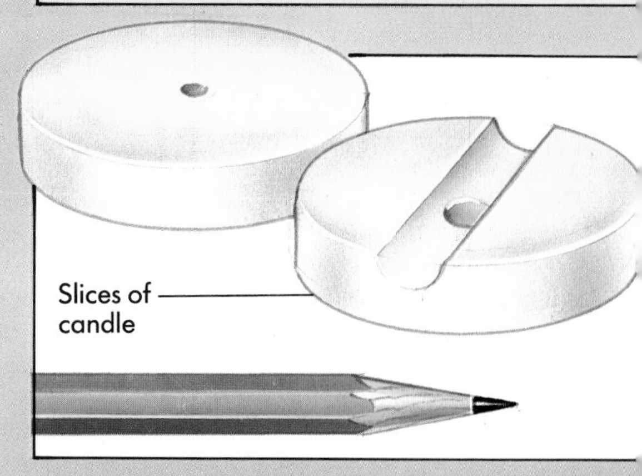

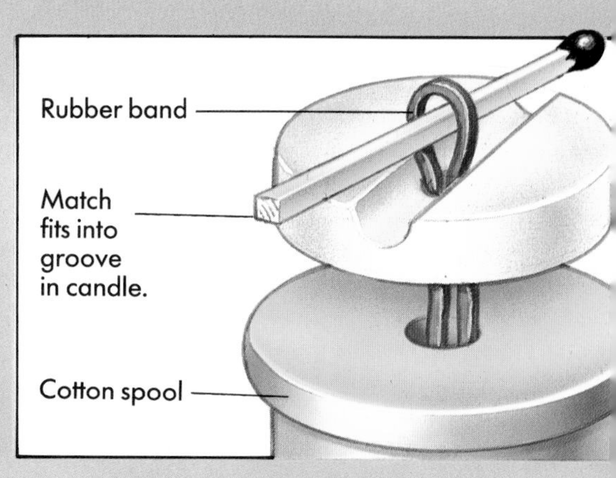

Dashing Darts

Equipment: A piece of balsa wood or thick polystyrene, two large hooks, two small hooks, a rubber band, beads, a short piece of wire, a propeller (from an old toy or a store that sells equipment to make models).

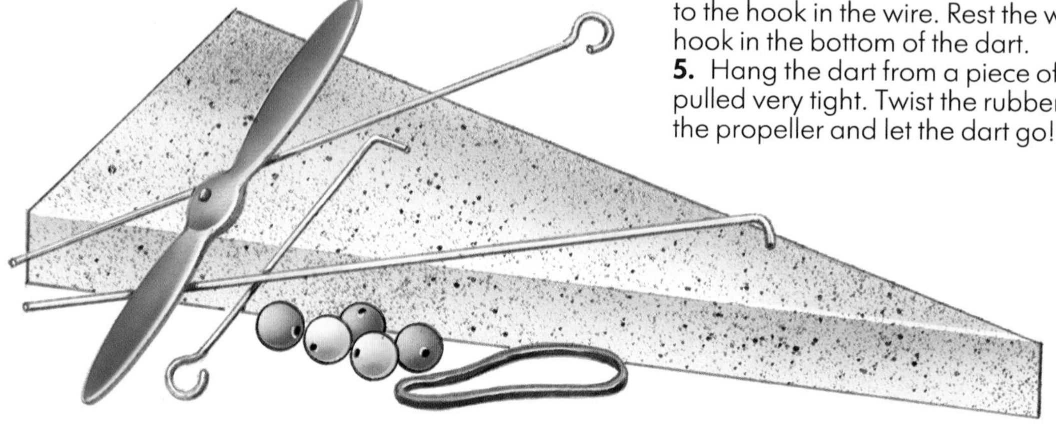

1. Shape the balsa wood or polystyrene to ma a dart shape.
2. Fix the two large hooks into the top of the da and two smaller hooks into the bottom.
3. Bend the wire to make a hook at one end. Thread the beads and the propeller over the otl end.
4. Attach one end of the rubber band to the fro hook in the bottom of the dart and the other enc to the hook in the wire. Rest the wire on the othe hook in the bottom of the dart.
5. Hang the dart from a piece of string that is pulled very tight. Twist the rubber band by turnii the propeller and let the dart go!

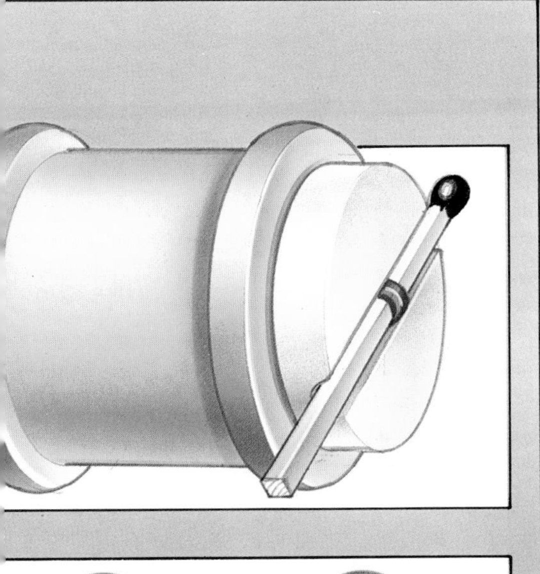

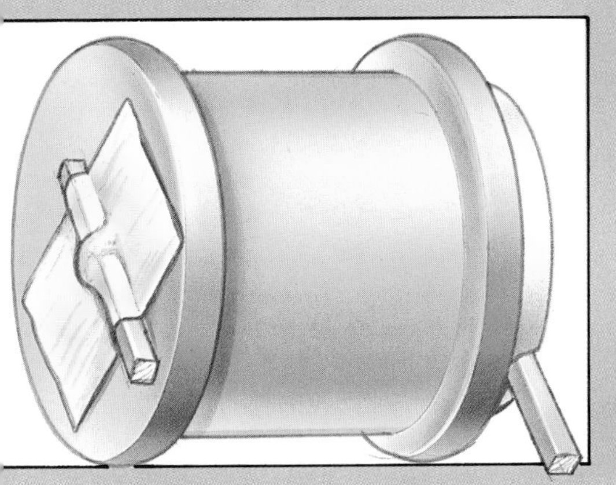

Make a Merry-go-round

Equipment: The same materials as for the creeping crawler plus a long, thin stick, some thin cardboard and a piece of thread.

Make the crawler toy as before but fix the thin stick at the candle end instead of the match. Make a small horse or plane out of the card and tie it on to the end of the thin stick using the thread. Turn the long stick several times to wind up the toy and then stand the cotton spool upright. You may find it easier to press the cotton spool into a piece of modeling clay to stop it wobbling.

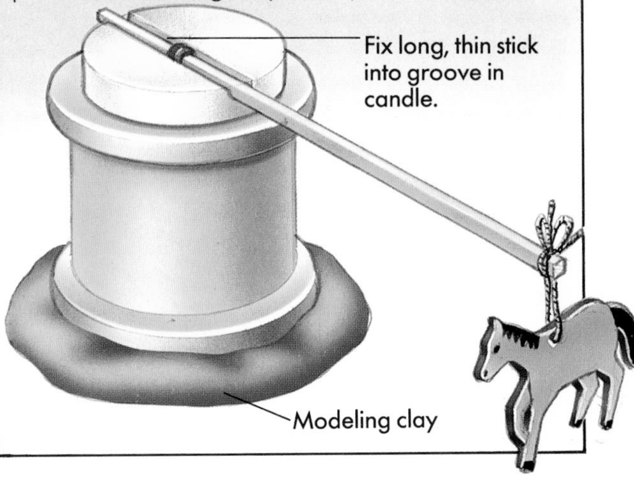

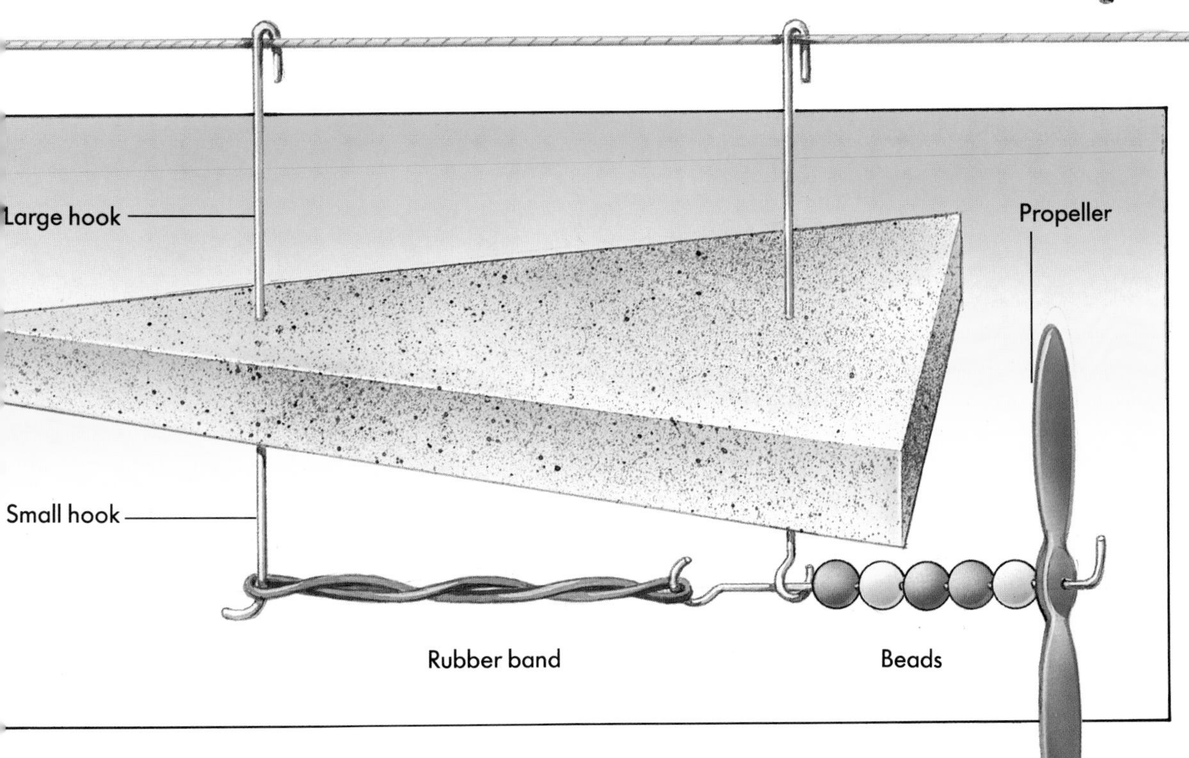

In a Spin

When an object spins around, it creates a force called **centrifugal force**, which pulls it outward. You can feel this force if you attach a piece of string to a ball and whirl it around and around in a circle. Centrifugal force is used in the machines at fairgrounds and to spin clothes dry. It even keeps satellites in orbit around the earth.

Pick up the Marble

Place a marble on a table and cover it with a glass jar. This trick will show you how to lift the marble without touching it.

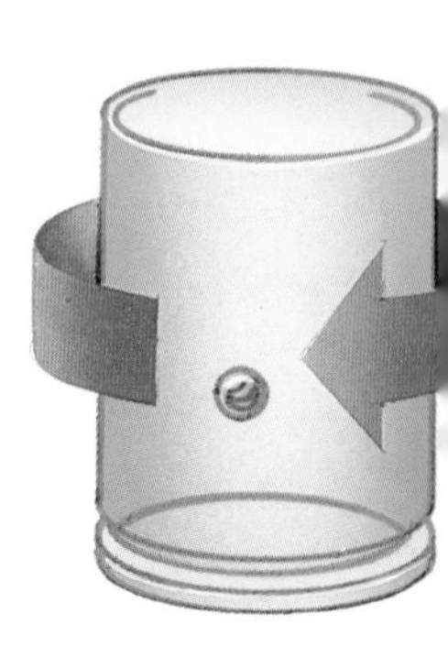

If you spin the jar around, this will start the mar spinning too. Eventually it will be pressed agai the sides of the jar by centrifugal force. The mo of the jar is narrower than the sides so the marb cannot fly out if you lift the jar.

▼ The "fly on the wall" ride at fairgrounds works using centrifugal force. The chamber in which the people stand spins around and centrifugal force presses them against the walls. The floor can be lowered leaving them "stuck" to the walls.

Spinning Water

›inning forces affect water too. Take a small
ıcket of water outside and try spinning it
ound in a circle quite quickly. Centrifugal
rce will keep the water pressed against the
›ttom and sides of the bucket while it spins
ound. This means the water will not fall out of
e bucket even when it is upside down!

Make a Spin Drier

Equipment: A hand drill or egg beater, thin string, yogurt pot, water, food coloring.

his investigation to see what happens to
er in a container that is spun horizontally and
out how this can be useful.

Make three small holes in the rim of the yogurt
and use the string to hang the pot from the
ing part of the drill or beater.
Put about 1 inch (3 cm) of water in the pot. You
ht like to color the water with a little food
oring so you can watch it more easily.
Turn the handle of the drill or beater steadily so
pot spins round and round. If you watch the
er, you will see it pulled against the sides of
pot by centrifugal force.

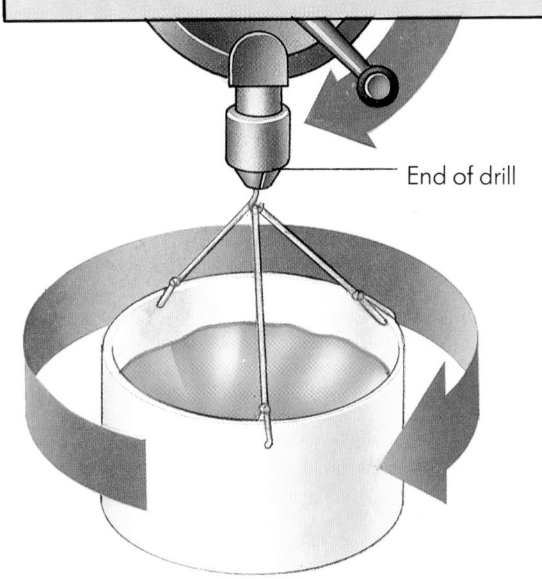

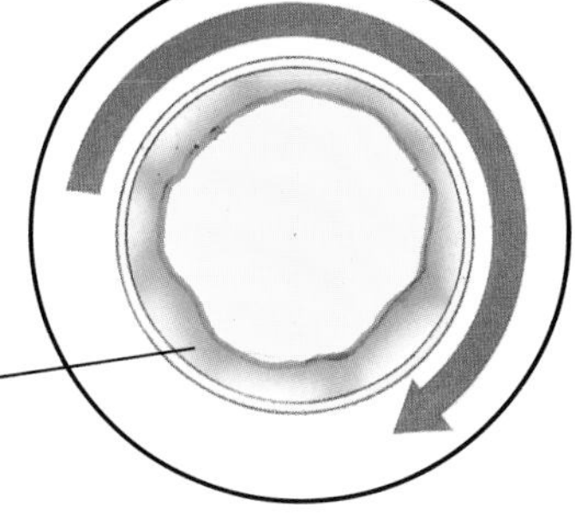

water is
hed against
sides of the

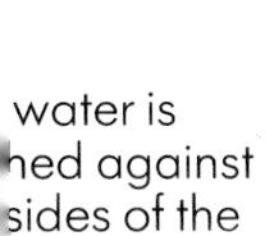

re things to try

ke holes in the sides of the yogurt pot and put
ery wet piece of cloth inside instead of the
er. Spin the pot outdoors or inside a very wide
ket or bowl. You will see that the water in the
h is thrown out of the pot by centrifugal force.
is just how a real spin drier works. The tub is
of holes and the water in the clothes is pushed
of the holes by centrifugal force as a motor
kes the tub spin around.

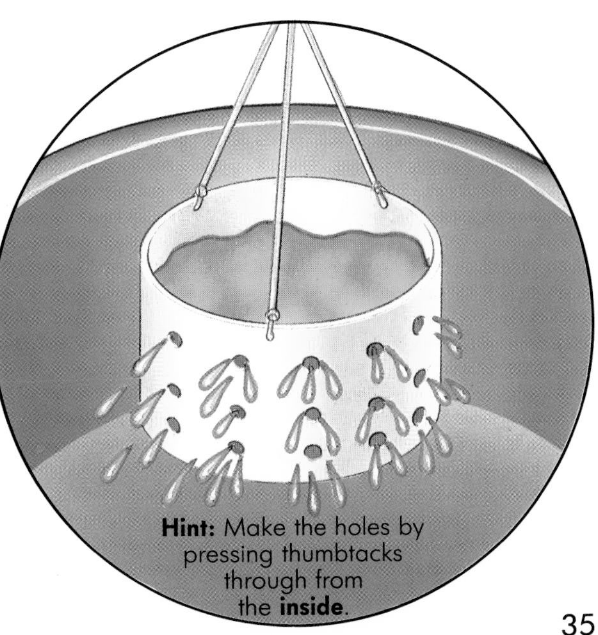

Hint: Make the holes by pressing thumbtacks through from the **inside**.

Machines and Movement

People use a variety of machines to make moving easier. The power to drive these machines comes from animals, which pull carts, plows, and sleds and also from natural forces, such as the wind and running water. Windmills and water mills have been used for thousands of years. Today most machines are driven by electricity, coal, or oil.

Windmills

People use the pushing power of the wind to drive machines. Windmills have been in use centuries and are still a common sight today some parts of the world, such as Holland an Greece. The blades of a windmill are shape to catch the wind and turn easily. As the blac turn, they make wheels move around to turn grindstones or make other machinery work. Today, scientists are trying out windmills designed to produce electricity.

► The blades on this modern experimental windmill are a special shape to help them ca the wind blowing from any direction.

Looking at Water Wheels

There are two main types of water wheel. **Undershot wheels** are moved around by the flo of water as it pushes against the flat paddles sticking out from the wheel.

Overshot wheels have bucket-shaped paddles catch the water. The weight of the water in the paddles helps to turn the wheel faster than the weight of the flowing water alone. This wheel needs a difference in the level of the water.

Make a Water Wheel

u can get a good idea of how a water wheel orks by making this model and using flowing ater from a faucet to make it turn around.

Cut four pieces of thin cardboard about 1½ ches × 1 inch (3½ cm × 2 cm).
Fold each "blade" in half and glue half of it ito the cotton spool.
Push the pencil or knitting needle through e hole in the middle of the cotton spool and ld it under a gently running faucet. The force the water will turn your "water wheel" ound.

Fold here

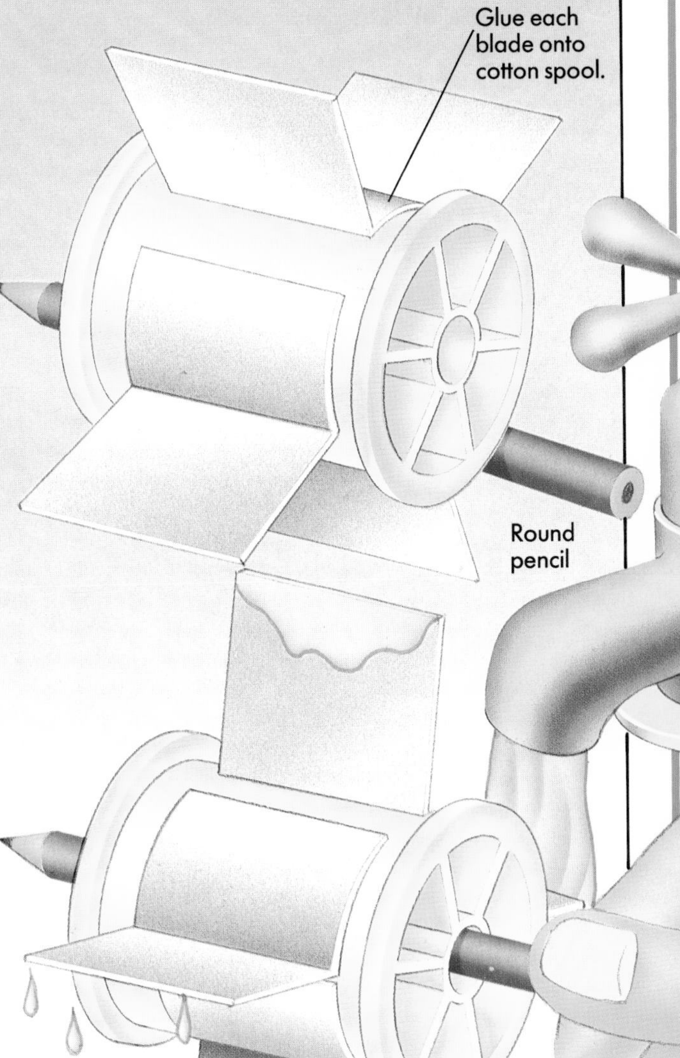
Glue each blade onto cotton spool.
Round pencil

Equipment: A cotton spool, a long, **round** pencil or knitting needle, cardboard, scissors, glue.

Steam Engines

The power of steam can also be used to make machines move. Over a hundred years ago, the first steam engines were being used to pull railroad cars and to drive the wheels that turned the machinery in mills and factories. Steam engines are still used in some parts of the world today but many have been replaced by engines powered by diesel or electricity.

▲ A steam train on the island of Java, Indonesia. Steam engines have to be kept filled up with water. They pull a trailer called a tender behind them to carry coal or wood for the fire and water to top up the boiler.

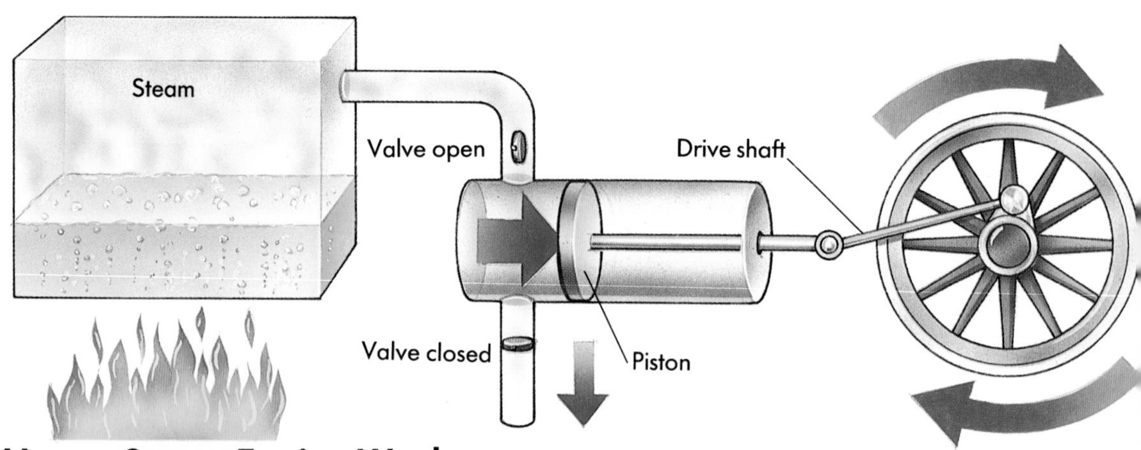

How a Steam Engine Works

A steam engine turns heat into mechanical energy which makes things move. The engine burns coal, oil, or wood to heat water until it turns into steam. The steam takes up more space than the water and pressure builds up.

The pressure of the steam pushes pistons inside metal cylinders. The pistons push a drive shaft, which turns a wheel. The wheel may either move the engine forward or it may turn other engine parts, which drive factory machinery.

Make a Steam Boat

Place the three candles inside the can.
Pour about ¾ inch (2 cm) of water into
e metal tube. Ask an adult to help you
ake a small hole in the screw cap of the
be and put the cap on the tube.
Use modeling clay to fix the metal tube
side the sardine can, over the candles.
Put the steam boat in the bath or in a
ool of water where there is plenty of space.
Light the candles and watch the boat
avel.

arning: Make sure you blow out the
atch and the candles
terward.

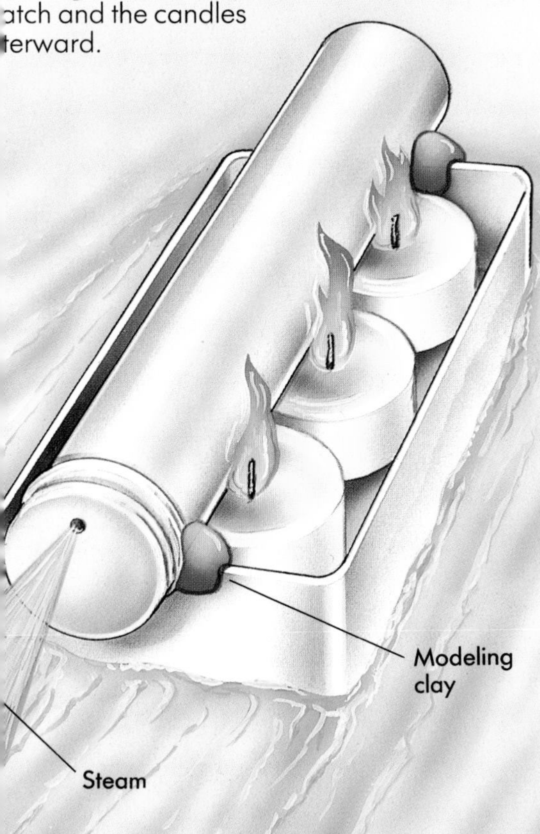

quipment: A clean sardine can, three thin
ces of candle, water, a hollow **metal** tube
uch as the one used to hold some
digestion tablets), modeling clay.

ow it works
s the candles heat the water in the tube, it
oils and turns to steam. The steam shoots
ut of the hole in the tube and pushes the
oat forward.

Turning Turbines

A turbine is a wheel that is turned by the force of water, steam, or gas. The wheel has hundreds of metal blades on a long axle. Turbines are used in power stations to provide the energy that turns the generators which produce electricity. They are also used to drive ships and submarines.

In about two-thirds of the world's power stations, strong jets of steam drive the turbine wheels around. In most other power stations, the power of flowing water is used to turn turbines and provide what is called "hydroelectric power" – "hydro" means "to do with water."

▼ Engineers testing a set of turbines before they are fitted into a power station. There are several turbines in the set, which helps to produce the maximum amount of energy from the steam.

Index

Editor: Barbara Taylor
Designer: Ben White
Illustrators: Kuo Kang Chen · Peter Bull
Consultant: Terry Cash

Additional Illustrations: Andrew Macdonald; pages 4, 5, 16, 17, 19 Catherine Constable; pages 28, 29, David Salariya and Shirley Willis; page 21
Cover Design: The Pinpoint Design Company
Picture Research: Jackie Cookson

Photograph Acknowledgements: 7 top Photri/ ZEFA; 11 bottom right Nik Cookson; 13 bottom Transport and Road Research Laboratory; 15 bottom Nik Cookson; 16 top ZEFA; 18 top Osh Kosh Truck Corp; 20 top J.Allan Cash; 24 left, 27 ZEFA; 28 bottom Argos Distributors Ltd; 34 bottom Blackpool Pleasure Beach, Astroswirl; 38 top ZEFA; 39 C.E.G.B.

Published 1987 by Warwick Press, 387 Park Avenue South, New York, New York 1001

First published in Great Britain by Kingfisher Books Limited.

Printed by South China Printing Company. H.K.
6 5 4 3 2 1

Library of Congress Catalog Card No. 86-51556
ISBN 0-531-19027-7